David Alkin

ELECTRONIC CIRCUIT ANALYSIS FOR SCIENTISTS

ELECTRONIC CIRCUIT ANALYSIS FOR SCIENTISTS

James A. McCray
Department of Biophysics
and Physical Biochemistry
University of Pennsylvania

Thomas A. Cahill
Department of Physics
University of California, Davis

JOHN WILEY & SONS
New York   London   Sydney   Toronto

This book is dedicated to our wives,
Diane McCray and Virginia Cahill

Copyright © 1973, by John Wiley & Sons, Inc.

All rights reserved. Published simultaneously in Canada.

No part of this book may be reproduced by any means, nor transmitted, nor translated into a machine language without the written permission of the publisher.

Library of Congress Cataloging in Publication Data

McCray, James A 1932–
   Electronic circuit analysis for scientists.

  Bibliography: p.
  1. Electronic circuits. 2. Electric networks.
I. Cahill, Thomas A., 1937– joint author.
II. Title.

TK7876.M3   621.3815'3   72-8986
ISBN 0-471-58295-6

Printed in the United States of America

10 9 8 7 6 5 4 3 2 1

PREFACE

When the fact that the majority of students studying physics, chemistry, and biology will eventually be engaged in experimental work is considered, it becomes evident that some early training in electronics is essential. However, the past attempts to satisfy this need have not produced the desired effect—namely the ability to quickly analyze, understand, and use electronic equipment intelligently and effectively. One of the basic problems, of course, is the limited amount of time that a student may devote to such a study.

Many good courses in electronics are offered in electrical engineering departments. The problem with these courses is that they are often part of a carefully designed program for electrical engineers and require extensive prerequisites. Also, they often involve material not really useful to students not planning a career in electronics.

One way around this problem is to offer courses in electronics as part of the curriculum in the physical or biological sciences. These courses unfortunately often become merely a survey of electronics, which, while introducing a student to a wide variety of circuits and techniques, never really allow him to do the calculations required by his particular research problems. Physicists have their own problems, in that electronics courses often become too involved in the physics of the real devices encountered which, although admittedly interesting, are not apt to help the student in a problem in the laboratory. Occasionally the courses provide a mechanism for introducing a little applied mathematics, which again may or may not help the student solve a circuit or connect two pieces of equipment together without engendering smoke.

It is necessary to resist the above temptations and design an electronics course which concentrates on circuit analysis and the practical use of electronic equipment. The physics of devices may be discussed outside the main lectures, in the laboratory, etc., and references can be given to the many excellent texts on physical electronics.

Such a course has been designed and given for the last eight years by the authors to students of physics, chemistry, biology, and applied science.

PREFACE

Usually the students were seniors or graduate students. The text is designed for and has been used in a one-quarter course with three lecture hours and one three-hour laboratory per week. The course has then been followed by a one-lecture hour, one three-hour laboratory course in the second quarter in which the students design, build, and operate their own circuits and electronics systems.

It has been our experience that most of the students enjoyed the course, and were no longer "afraid" of electronics. Those students going on to graduate school have demonstrated a definite ability to work with electronics, while those terminating at the Bachelor's level find their background in electronics to be a real asset in obtaining a position.

What we have tried to do in this text is to bridge the gap between the more sophisticated science student and the electronics engineer. By "sophisticated" we mean having some familiarity with calculus, some notion of ordinary differential equations, and the kind of physics or chemistry background offered by a reasonable introductory course. We hope that students who have studied this text will be able to read the vast electronics literature available and will be able to communicate profitably with the electronics people with whom they will come into contact.

The authors would like to acknowledge the important role which our students have played in the development of the text. Particular mention must be made of the teaching assistants, Steve Brooks, Gary Smith, Don McCauley, Stan Johnson, Bob Eldred, and Mike Ellison, especially with regard to the laboratories. Another student, George Ellis, drafted all the diagrams and waveforms for the text. We would also like to thank Ginny Cahill, whose extensive editorial assistance helped to prepare the manuscript for publication. Finally, we would like to acknowledge the support of the editors of John Wiley & Sons, especially Gary Brahms and Don Deneck, whose help and encouragement made this book possible.

James A. McCray
Thomas A. Cahill

CONTENTS

1 ANALYSIS OF PASSIVE ELEMENTS DIFFERENTIAL EQUATIONS KIRCHHOFF'S LAWS — 1

- Introduction — 1
- Kirchhoff's Laws — 2
- Differentiating and Integrating Circuits — 5
- Sinusoidal Input to Simple Circuits — 13
- The Uncompensated Attenuator — 17
- The Compensated Attenuator — 22
- RCL Circuits — 26
- Phasors and Complex Impedance — 30
- Q of a Circuit — 35
- Appendix — 36
- Problems — 38
- Experiment 1 (RC and RL Circuits) — 40

2 ANALYSIS OF PASSIVE ELEMENTS—LAPLACE TRANSFORMS — 45

- The Laplace Transform — 45
- The DC Circuit Scheme — 50
- Transformers — 57
- RF Circuitry — 63
- More about the Transfer Function — 64
- Appendix — 66
- Problems — 67
- Experiment 2 (RCL Circuits) — 69
- Appendix to Experiment 2 (Lissajous Figures) — 72

3 TRANSMISSION AND DELAY LINES — 75

- The Loss-less Transmission Line — 75
- Termination — 79

CONTENTS

Attenuation	90
Non-Inductive Transmission Line	91
RF Transmission Lines	94
NMR Probes	97
Problems	101
Experiment 3 (Transmission Lines)	104

4 VOLTAGE AMPLIFIER CIRCUITS—TUBES AND FETS — 107

Think Voltage	107
High-Input-Impedance Devices	110
Linear, or Small-Signal, Analysis	115
Time Response of Triode and FET Circuits	120
Multistage Amplifiers	127
Improvements of the Triode and FET	134
Experiment 4 (Triode Amplifiers)	136
Experiment 5 (FET Amplifiers)	142

5 FEEDBACK — 147

Introduction	147
Negative Voltage Feedback	149
Cathode Follower	152
Two-Stage Negative Voltage Feedback Amplifier	156
Positive Voltage Feedback	159
Wien Bridge Oscillator	160
Problems	163
Experiment 6 (Two-Stage Pentode Negative Voltage Feedback Amplifier)	165
Experiment 7 (FET Amplifiers with Feedback)	169

6 CURRENT AMPLIFIER CIRCUITS—TRANSISTORS — 171

Think Current	171
Transistors	172
Transistor Equations	174
Bias Conditions and Stabilization	176
Time Response of Transistor Circuits; Common-Emitter Mode of Operation	180
Common-Base Mode of Operation	184
Appendix	187
Problems	188
Experiment 8 (Bias Conditions and Gain of a Transistor Amplifier)	189

CONTENTS

7 AMPLIFIERS WITH SEVERAL ACTIVE ELEMENTS — 191

Introduction — 191
The Darlington Circuit — 192
A Difference Amplifier — 196
Rush Transistor Current Amplifier — 199
Cascaded Amplifiers — 204
Gaussian Amplifier — 205
Experiments 9 and 10 (Darlington Circuit and Difference Amplifier) — 207

8 OPERATIONAL AMPLIFIERS — 209

The Op-amp Configuration — 209
Concept of the Virtual Ground — 211
Operational Configurations for Differentiation, Integration, etc. — 212
A Charge-Sensitive Preamplifier — 215
An Active Delay-Line Differentiator — 217
Linear Integrated Circuits — 219
Logarithmic Amplifiers — 220
Problems — 222
Special Project in Operational Amplifier Analysis — 224
Experiment 11 (Operational Amplifiers and Feedback) — 225

9 NOISE — 235

Sources of Noise — 235
Signal-to-Noise Ratio — 238
Bandwidth and Noise — 239
CAT and Lock-in Amplifiers — 244

10 NONLINEAR CIRCUIT ANALYSIS—DIODES — 247

Nonlinear Systems — 247
Thermionic and Junction Diodes — 248
The Piecewise-Continuous Model — 250
Limiters and Clippers — 253
Clamping Circuits — 255
Diode Logic — 257
Rectifiers, Zener Diodes, and Power Supplies — 259
Tunnel Diodes — 265
Problems — 266
Experiment 12 (Diodes as Clippers and Clampers) — 267

CONTENTS

11 TRIGGERS AND MULTIVIBRATORS — 271

Large-Signal Transistor Analysis — 271
Multivibrators — 273
The Bistable Multivibrator (or Binary) — 273
The Schmitt Trigger — 279
The Monostable Multivibrator — 281
The Astable Multivibrator — 282
Tunnel Diode Multivibrators and Discriminators — 283
Experiment 13 (The Transistor Binary) — 286

BIBLIOGRAPHY — 288

INDEX — 289

NOMENCLATURE

(1) V, I, Q, P — These quantities are constant or direct current (dc) values except in situations in which an alternating current (ac) is superimposed on top of a direct current. In these cases, they are total values for voltage, current, charge, and power.

(2) v, i, q, p — These quantities represent deviations from dc values, usually in time. $dV/dt = d(V_{\text{DC}} + \Delta V)/dt = dv/dt$. Equations in these variables are called "ac" or "signal" equations.

(3) $\bar{V}, \bar{I}, \bar{Q}, \bar{P}$ — These quantities always represent the Laplace transforms of v, i, q, or p, and may or may not be written explicitly as $\bar{V}(s)$, $\bar{I}(s)$, etc.

(4) — This symbol represents an ideal voltage source, one whose internal impedance is zero.

(5) — This symbol represents an ideal current source, one whose internal impedance is infinite.

(6) $v(t)$ — This nomenclature merely indicates the voltage between the two points.

(7) Q — Q is unfortunately used universally for four quantities: 1. Charge. 2. "Q" of a circuit. 3. Operating point Q. 4. Transistor label Q.

ANALYSIS OF PASSIVE ELEMENTS
DIFFERENTIAL EQUATIONS
KIRCHHOFF'S LAWS

INTRODUCTION

To have a working model of electronics, two problems must be surmounted. First, real devices and circuits are so complicated in an exact analysis that it would be impossible to treat all but the simplest cases rigorously. Therefore, basic models must be assumed for the various components of a real circuit, and these models must be justified on the basis of simplicity and ability to predict that which is observed in the laboratory. Second, once models are assumed for the components, methods must be developed to solve the integro-differential equations that arise when the elements are combined in circuits and subjected to realistic input conditions. A third point might also be raised: that in one book, or even a set of books, only a small fraction of

PASSIVE ELEMENTS: DIFFERENTIAL EQUATIONS

the vast field of electronics can be covered. We feel strongly that the inevitable selection of topics should favor those that can help a scientist in the laboratory make best use of the tremendous resources of electronics.

To handle the first problem, we have used a *lumped parameter model* for passive elements; resistance, capacitance, and inductance. When this model is insufficient, as in the case of RF transmission lines, modifications are made that cover the deviations. For active elements, we have used a linear approximation for the first nine chapters, and have divided active elements into those possessing high input impedance (tubes, FET's, etc.) and low input impedance (transistors). This division introduces simplifications in the analysis and leads naturally to voltage and current devices.

To handle the second problem, we have not relied upon previous knowledge of differential equations beyond a basic familiarity with integrals and differentials. This is not enough, for as circuits become realistic, the equations become horrendous. Therefore, as early as possible in the second chapter, we introduce *Laplace transforms*, which reduce coupled integro-differential equations to simple algebra. We then use the *dc Circuit Scheme* with the Laplace transforms, and the result is a method that allows a scientist to solve real circuits and find output voltages without ever setting up the coupled equations. We do this by avoiding the temptation to do contour integrals (which we admit is fun) by using tables of inverse transforms. Finally, we show that by using *pole-zero plots*, even the inverse transform tables can be avoided in many occasions. We have found that our students have thus gained both a real facility with electronics and an understanding of circuits that would be extremely hard to gain any other way. These methods have, of course, been known to electrical engineers for some time, but they have not been widely used on the level of a physical or biological scientist.

To handle the third problem, we have selected topics both by their utility to a scientist in a modern laboratory and by their contribution to an understanding of circuits. To maintain this book within limits set by a single quarter of instruction, we have deleted almost all discussion of the physics of devices. All that is left is what we feel is a bare minimum required for appreciation of the simple models that we use. However, many good books exist on all levels of sophistication for the physical analysis of real devices, and we refer to these, listed in the Bibliography, for the details. Also, this material can be added by the instructor in his lectures or in the laboratory at exactly the level that suits his class.

KIRCHHOFF'S LAWS

As we have discussed in the introduction, we will use the lumped parameter model for the passive elements: resistance, capacitance, and inductance.

KIRCHHOFF'S LAWS

We base our model on the following results, which are assumed valid for ideal lumped parameter components:

1. The energy dissipated in a resistor is

$$W_R = R \int_0^t i^2 \, dt,$$

where R is the *resistance* and i is the *current* passing through the resistor.

2. The energy stored in a capacitor (potential energy) is

$$W_C = \frac{1}{2}\frac{q^2}{C},$$

where C is the *capacitance* and q is the *charge* on the capacitor.

3. The energy stored in an inductor ("kinetic energy") is

$$W_L = \tfrac{1}{2}Li^2,$$

where L is the *inductance*.

The physical laws which apply to electronic circuits are the *conservation of charge* and the *conservation of energy*. The first conservation law leads to Kirchhoff's first law, which may be written

> *At a node*
> Σ (current in) $= \Sigma$ (current out).

The second conservation law leads to Kirchhoff's second law, which may be written

> *Around a closed loop*
> Σ (voltage sources) $= \Sigma$ (voltage drops).

To illustrate the second conservation law, we consider the single loop circuit of Fig. 1-1.

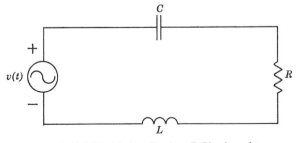

FIGURE 1-1. Series *RCL* circuit.

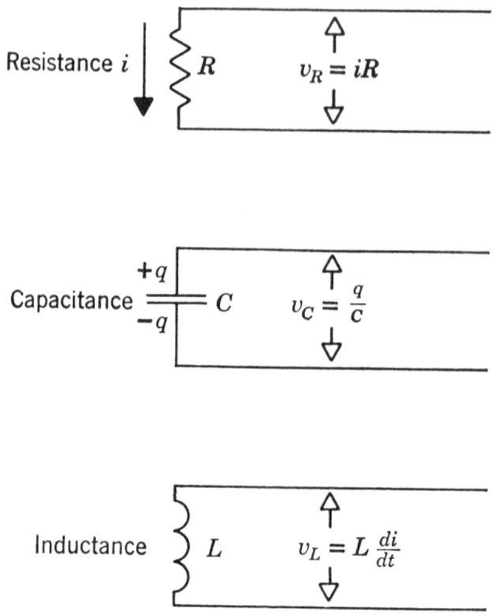

PASSIVE ELEMENTS: DIFFERENTIAL EQUATIONS

FIGURE 1-2. Voltage drop with resistive, capacitive, and inductive elements.

The energy put into the circuit from the voltage source is given by

$$W = \int_0^q v(t)\, dq = \int_0^t vi\, dt.$$

Conservation of energy requires that

$$W = W_R + W_C + W_L,$$

or

$$\int_0^t vi\, dt = \int_0^t i^2 R\, dt + \frac{q^2}{2C} + \frac{L}{2} i^2.$$

If we differentiate this equation with respect to time and cancel a common factor i, we have

$$v(t) = iR + \frac{q}{C} + L\frac{di}{dt},$$

or

$$v(t) - L\frac{di}{dt} = \frac{q}{C} + iR.$$

$$\begin{pmatrix}\text{External}\\ \text{voltage}\\ \text{source,}\\ \text{or e.m.f.}\end{pmatrix} + \begin{pmatrix}\text{Back}\\ \text{e.m.f.}\\ \text{due to}\\ \text{inductance}\end{pmatrix} = \begin{pmatrix}\text{Voltage}\\ \text{across}\\ \text{capacitor}\end{pmatrix} + \begin{pmatrix}iR \text{ drop}\\ \text{across resistor}\\ \text{(Ohm's law)}\end{pmatrix}.$$

DIFFERENTIATING CIRCUIT

We shall work, then, with passive elements having the "voltage drops" shown in Fig. 1-2.

DIFFERENTIATING AND INTEGRATING CIRCUITS

Now let us consider some very simple but useful passive circuits. The first circuit to be considered is the *differentiating circuit;* however, it is also known by such other names as blocking circuit, clipping circuit, high-pass circuit, and lead circuit. This copious supply of names brings out a very important point about electronic circuits. A given configuration of circuit elements may have several functions depending upon the relative values of the circuit parameters and the parameters of the input wave form. Throughout this book, $v_0(t)$ is assumed to be an ideal voltage source; that is, one having zero internal resistance.

In analyzing the circuit of Fig. 1-3, we first assume that the load current $i_2(t)$ is very small compared with the loop current $i(t)$. Application of Kirchhoff's second law yields

$$v_0(t) - \frac{q(t)}{C} - i(t)R = 0,$$

which may be written as a differential equation for $q(t)$:

$$\frac{dq}{dt} + \frac{1}{RC} q = \frac{v_0(t)}{R}.$$

Let us initially solve the above equation by standard techniques of differential equations.* The general solution is the sum of the homogeneous solution and the particular solution:

$$q(t) = q_h(t) + q_p(t).$$

The particular solution will yield the long-term (steady-state) response to an input while the homogeneous solution will give the decaying (transient) behavior.

For the homogeneous part, we have

$$\frac{dq_h}{dt} + \frac{q_h}{RC} = 0,$$

with solution

$$q_h(t) = Ae^{-t/RC}$$

where A is a constant determined by initial conditions.

---

* See any good reference to differential equations, such as I. S. Sokolnikoff and R. M. Redheffer, *Mathematics of Physics and Modern Engineering*, McGraw-Hill, New York (1966), Chapters 2 and 3.

PASSIVE ELEMENTS: DIFFERENTIAL EQUATIONS

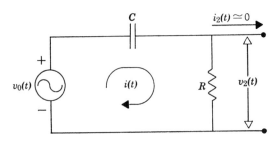

FIGURE 1-3. RC differentiating circuit.

The particular solution may be easily found by the *method of undetermined coefficients*, which applies when the right-hand side of the differential equation contains only terms from which a finite number of terms can be obtained by differentiation. We use as a trial function for the particular solution, the inhomogeneous term, plus all of its derivatives. We find it useful to obtain the response $v_2(t)$ of a given circuit to a step input $v_0(t) = Vu(t)$, where

$$u(t) = \begin{cases} 1 & 0 < t \\ 0 & t \leq 0 \end{cases}$$

is the *unit step function*, and to a sinusoidal input $v_0(t) = V \sin(\omega t + \phi)$, where V is the *amplitude*, ω the angular *frequency*, and ϕ the *phase* of the input sinusoidal waveform. The response to the step input tells us how fast a circuit can respond to a sudden discontinuity in voltage, and the response to a sinusoidal input gives us the circuit's steady-state characteristics.

For a step input, the differential equation is

$$\frac{dq}{dt} + \frac{q}{RC} = \frac{V}{R}.$$

For the particular solution, we have as a trial function

$$q_p(t) = B \text{ (a constant)}.$$

Substitution of this trial solution into the differential equation yields the value of $B = CV$. The solution is

$$q(t) = Ae^{-t/RC} + CV.$$

We now determine A by the initial condition $q(0) = q_0$. Then

$$q(t) = (q_0 - CV)e^{-t/RC} + CV.$$

The response of the differentiating circuit to a step input is

$$v_2(t) = i(t)R.$$

DIFFERENTIATING CIRCUIT

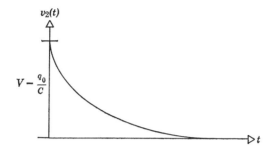

FIGURE 1-4. Output of an RC differentiating circuit for an input step function.

Or since $i(t) = dq/dt$,

$$v_2(t) = \left(V - \frac{q_0}{C}\right)e^{-t/\tau},$$

where $\tau = RC$ is the *time constant* of the circuit, the most important circuit parameter. For the case we are considering (Fig. 1-4), the signal falls to about $\frac{1}{3}$ of its initial value in time τ. This is characteristic of an exponential decay.

Actually, we are more interested in the response of the circuit to a pulse, so let us take for $v_0(t)$ the *rectangular pulse* of height V and width T_0 (Fig. 1-5). We may solve this problem by considering different time intervals. Let us assume that initially there is no charge on the capacitor. We have the solution for time interval I:

$$v_2(t) = Ve^{-t/\tau} \qquad 0 \leq t \leq T_0.$$

In time interval II, the differential equation is

$$\frac{dq}{dt'} + \frac{1}{\tau}q = 0,$$

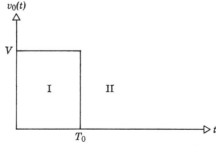

FIGURE 1-5. Rectangular pulse.

7

PASSIVE ELEMENTS: DIFFERENTIAL EQUATIONS

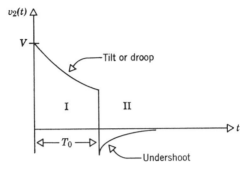

FIGURE I-6. Output of an RC differentiating circuit for an input rectangular pulse.

where $t' = t - T_0$. This has the solution

$$q(t') = Ae^{-t'/\tau}.$$

The initial condition now is

$$q(t') = q(t - T_0) = q(0) = CV(1 - e^{-T_0/\tau}) = A.$$

Hence,

$$q(t) = CV(1 - e^{-T_0/\tau})e^{-(t-T_0)/\tau}$$

and

$$v(t) = i(t)R = -V(1 - e^{-T_0/\tau})e^{-(t-T_0)/\tau} \qquad T_0 \leq t.$$

The response is as shown in Fig. 1-6.

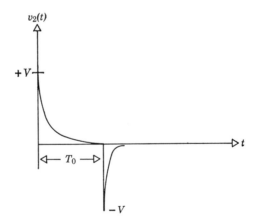

FIGURE I-7. Output of an RC differentiating circuit for an input rectangular pulse: $\tau \ll T_0$.

DIFFERENTIATING CIRCUIT

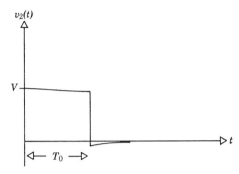

FIGURE 1-8. Output of an *RC* differentiating circuit for an input rectangular pulse: $\tau \gg T_0$.

We see that the circuit shapes the pulse and that the shaping depends upon the relative value of the decay time constant of the circuit and the width of the pulse. In particular, if $\tau \ll T_0$ (Fig. 1-7), spikes are produced. If $\tau \gg T_0$ (Fig. 1-8), the pulse shape is not changed very much. This situation occurs in *RC* coupling between amplifier stages when it is necessary to block a dc voltage. It is undesirable there to have the coupling circuit shape the pulse, so the time constant is made very large, giving a *blocking circuit*.

For very small time constants (small compared with the time necessary for the input to change significantly), the capacitor will charge rapidly, and after a very short time the voltage across R will be small compared to that across C.

Then

and
$$\frac{q}{C} = v_C \simeq v_0$$

or
$$v_2(t) = i(t)R = \frac{dq}{dt} R,$$

$$v_2(t) \simeq \tau \frac{dv_0}{dt}.$$

Thus, after a very short time the output voltage is proportional to the derivative of the input voltage. For this reason, the circuit is also called a *differentiating circuit*.

The second circuit we wish to consider is the *integrating circuit*, known also as the smoothing circuit, delay circuit, low-pass circuit, and lag circuit (Fig. 1-9).

PASSIVE ELEMENTS: DIFFERENTIAL EQUATIONS

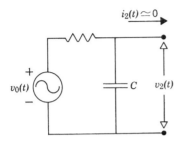

FIGURE 1-9. RC integrating circuit.

Again we assume infinite load conditions and apply Kirchhoff's laws. We have exactly the same differential equation as before, only now

$$v_2(t) = \frac{q(t)}{C}.$$

Thus,

$$v_2(t) = \frac{q_0}{C} e^{-t/\tau} + V(1 - e^{-t/\tau})$$

is the response of the circuit to a voltage step input.

In working with electronic circuits, we are usually interested only in deviations from quiescent or equilibrium values of charge, current and voltage. Hence, we will be working with first-order deviations; that is, with variables which will automatically have zero initial conditions. The linear analysis is thus greatly simplified, and we will usually consider cases of zero initial conditions. For the integrating circuit, we would have the exponential rise, or charging function, shown in Fig. 1-10.

The above circuit turns out to be a very good equivalent circuit for the short time response of many electronic circuits, and its response may be used

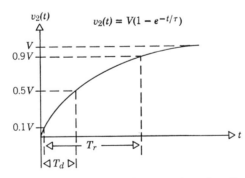

FIGURE 1-10. Exponential rise, or charging function.

INTEGRATING CIRCUIT

to define some very important parameters of such circuits. The most important parameter is the *rise time* of the circuit. There are several definitions but the most useful is the 10–90% rise time. This is defined to be the time necessary for the output voltage to rise from 10% to 90% of its final value and is found very easily from an oscilloscope trace. For the circuit above, the rise time is simply related to the time constant of the circuit:

$$T_r \cong 2.2\tau.$$

As we shall see, the expected rise time of a circuit may be estimated from the circuit diagram.

The other important parameter of this circuit is the *delay time* T_d. This parameter is defined to be the time necessary for the output voltage to rise to one half of its final value and becomes especially important when one is working with time coincidence experiments. For the above circuit, the delay time may also be related to the time constant:

$$T_d = 0.693\tau.$$

The response to a rectangular pulse is now easily found. Again, we assume an uncharged capacitor initially. For time interval I, we have

$$v_2(t) = V(1 - e^{-t/\tau}) \qquad 0 \leq t \leq T_0.$$

In time interval II, with the input at zero voltage, it is effectively shorted so that we have a simple RC circuit which has now an initial charge of

$$q(t' = 0) = CV(1 - e^{-T_0/\tau}).$$

This charge, then, simply decays to give us

$$v_2(t) = V(1 - e^{-T_0/\tau})e^{-(t-T_0)/\tau} \qquad T_0 \leq t.$$

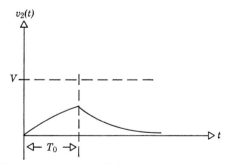

FIGURE I-11. Output of an RC integrating circuit for an input rectangular pulse.

PASSIVE ELEMENTS: DIFFERENTIAL EQUATIONS

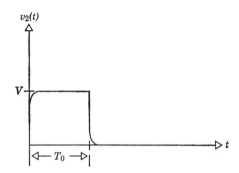

FIGURE 1-12. Output of an *RC* integrating circuit for an input rectangular pulse: $\tau \ll T_0$.

Thus, the response is that the output wave form is drastically modified (Fig. 1-11). The *relative values* of the *pulse width* and *time constant* of the circuit are *extremely important*. For the limiting case of short time constant ($\tau \ll T_0$) (Fig. 1-12), the shape and magnitude of the pulse are hardly changed at all. However, if $\tau \gg T_0$ (Fig. 1-13), then we almost lose the pulse entirely. This becomes a very important factor when one is interested in pulse height analysis. Although the relative values of different pulses will be maintained for an ideal circuit of the type above, this situation of large time constant for this integrating circuit essentially "undoes" ones' previous efforts to amplify a pulse and, hence, is usually undesirable. The circuit has too long a rise time for pulses of this width. One must obtain a circuit with shorter rise time if these pulses are to be faithfully reproduced.

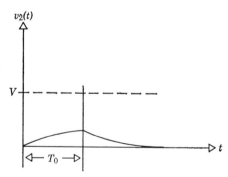

FIGURE 1-13. Output of an *RC* integrating circuit for an input rectangular pulse: $\tau \gg T_0$.

SINUSOIDAL INPUT

SINUSOIDAL INPUT TO SIMPLE CIRCUITS

Now let us determine the response of the simple circuit (Fig. 1-14) to a sinusoidal input. The differential equation becomes

$$\frac{dq}{dt} + \frac{1}{\tau} q = \frac{V}{R} \sin(\omega t + \phi).$$

The homogeneous solution is again a simple exponential decay:

$$q_h = A e^{-t/\tau}.$$

We find the particular solution by taking as a trial solution the inhomogeneous term and all terms obtained from it by differentiation. This would then be a sum of sines and cosines since

$$q_p = B \sin(\omega t + \phi) + D \cos(\omega t + \phi)$$

and

$$\frac{dq_p}{dt} = \omega B \cos(\omega t + \phi) - \omega D \sin(\omega t + \phi).$$

Substituting into the above equation yields

$$\left(\omega B + \frac{D}{\tau}\right) \cos(\omega t + \phi) + \left(\frac{B}{\tau} - \omega D\right) \sin(\omega t + \phi) = \frac{V}{R} \sin(\omega t + \phi).$$

It this equation is to be true for all times, the coefficients of like sinusoidal functions must be equal. Hence, we obtain

$$D = -B\omega\tau$$

and

$$B = \frac{CV}{1 + (\omega\tau)^2}.$$

The particular solution thus becomes

$$q_p(t) = \frac{CV}{1 + (\omega\tau)^2} [\sin(\omega t + \phi) - \omega\tau \cos(\omega t + \phi)].$$

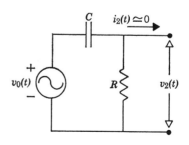

FIGURE 1-14. RC differentiating circuit.

13

PASSIVE ELEMENTS: DIFFERENTIAL EQUATIONS

We may simplify this by using the following trigonometric relation:

$$A \sin \theta + B \cos \theta \equiv \sqrt{A^2 + B^2} \sin\left(\theta + \tan^{-1} \frac{B}{A}\right).$$

Hence, we have

$$q_p(t) = \frac{CV}{[1 + (\omega\tau)^2]^{1/2}} \sin(\omega t + \phi - \tan^{-1} \omega\tau).$$

Now let us assume initial conditions $q = q_0$ at $t = 0$. Then

$$A = q_0 - \frac{CV}{[1 + (\omega\tau)^2]^{1/2}} \sin(\phi - \tan^{-1} \omega\tau).$$

The final solution is

$$q(t) = \left\{q_0 - \frac{CV \sin(\phi - \tan^{-1} \omega\tau)}{[1 + (\omega\tau)^2]^{1/2}}\right\} e^{-t/\tau}$$

$$+ \frac{CV}{[1 + (\omega\tau)^2]^{1/2}} \sin(\omega t + \phi - \tan^{-1} \omega\tau)$$

and consists of two separate parts. The first part is called the *transient solution* and decays exponentially with time. The second part is called the *steady-state solution* and is the only surviving solution after an appreciable time has elapsed.

Now let us determine the response of the differentiating circuit. We differentiate the above charge equation to find the current. Ohm's law then immediately gives us the output voltage:

$$v_2(t) = -\left\{\frac{q_0}{C} - \frac{V}{[1 + (\omega\tau)^2]^{1/2}} \sin(\phi - \tan^{-1} \omega\tau)\right\} e^{-t/\tau}$$

$$+ \frac{\omega\tau V}{1 + (\omega\tau)^2]^{1/2}} \cos(\omega t + \phi - \tan^{-1} \omega\tau).$$

Let us now consider only the steady-state solution. The "gain" of the circuit is then defined by

$$G(\omega) = \frac{|v_2(t)|}{V} = \frac{\omega\tau}{[1 + (\omega\tau)^2]^{1/2}}.$$

For very low frequencies, this is small. For $\omega\tau \gg 1$, this approaches unity. Hence, the circuit is a high-pass circuit; that is, only higher frequencies are passed without distortion (Fig. 1-15).

SINUSOIDAL INPUT

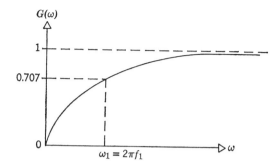

FIGURE 1-15. Gain of an *RC* differentiating circuit.

Now let us consider the phase response of the differentiating circuit (Fig. 1-16). We write $\cos(\omega t + \phi - \tan^{-1} \omega\tau)$ as $\sin(\omega t + \phi + \pi/2 - \tan^{-1} \omega\tau)$ in order to compare the phase with that of the original signal $\sin(\omega t + \phi)$. We see that there is a phase shift:

$$\Delta\phi = \frac{\pi}{2} - \tan^{-1} \omega\tau.$$

At low frequencies, the output voltage leads the input by about 90°. At high frequencies, $\tan^{-1}\omega\tau \to \pi/2$, so $\Delta\phi \to 0$ and there is no phase shift.

At the frequency corresponding to the condition $\omega\tau = 1$, the gain is

$$G(\omega_1) = \frac{1}{\sqrt{2}} = 0.707.$$

The frequency $f_1 = 1/2\pi\tau$ is called the lower half-power point, or lower 3-dB point. This latter terminology comes from the idea that power is proportional to voltage squared. So if the power is down by a factor of 2, the voltage (or gain) is down by a factor of square root of 2. The power ratio

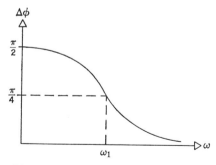

FIGURE 1-16. Phase response of an *RC* differentiating circuit.

PASSIVE ELEMENTS: DIFFERENTIAL EQUATIONS

is quite often given in terms of the decibel, defined in the following way:

$$N = \text{number of decibels} = 10 \log_{10} \frac{P}{P_{max}}.$$

Hence, for

$$\frac{P_1}{P_{max}} = \frac{1}{2},$$

we have

$$N = 10 \log_{10} \tfrac{1}{2} = -10 \log_{10} 2 \simeq -10(0.300),$$

or

$$N \simeq -3.$$

We say the gain, or response, is down 3 dB.

We consider now the integrating circuit. As indicated above, we usually ignore the transient solution when considering the response to sinusoidal inputs and consider only the steady-state solutions. The output voltage may be written in the following way:

$$v_2(t) = \frac{q_2(t)}{C} = \frac{V}{[1 + (\omega\tau)^2]^{1/2}} \sin(\omega t + \phi - \tan^{-1} \omega\tau).$$

Hence, the gain is given by

$$G(\omega) = \frac{1}{[1 + (\omega\tau)^2]^{1/2}}.$$

We see from Fig. 1-17 that only lower frequencies are passed by the circuit. Hence, this circuit is known also as a low-pass circuit. At the frequency corresponding to the condition $\omega\tau = 1$, the gain of this circuit is also

$$G(\omega_2) = \frac{1}{\sqrt{2}} = 0.707,$$

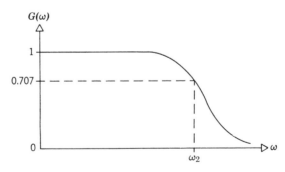

FIGURE 1-17. Gain of an *RC* integrating circuit.

THE UNCOMPENSATED ATTENUATOR

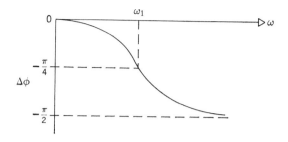

FIGURE 1-18. Phase response of an *RC* integrating circuit.

and the frequency $f_2 = 1/2\pi\tau$ is called the upper half-power point, or upper 3-dB point. The phase shift may be determined at once to be

$$\Delta\phi = -\tan^{-1}\omega\tau.$$

Hence, at low frequencies, the output is approximately in phase with the input; while for very high frequencies, the phase shift approaches $-90°$; the output lags the input (Fig. 1-18).

We have seen how simple *RC* circuits can perform the mathematical operations of differentiation and integration, to some degree of approximation. An even simpler operation is performed by the *attenuator* (Fig. 1-19), which divides an input voltage by a constant factor, or

$$v_2(t) = \frac{R_2}{R_1 + R_2} v_0(t).$$

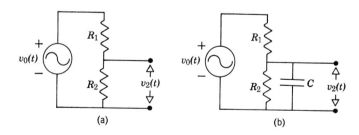

FIGURE 1-19. Uncompensated attenuator, without and with input capacitance in the load.

PASSIVE ELEMENTS: DIFFERENTIAL EQUATIONS

A common example of an attenuator is an oscilloscope probe, which is useful when one encounters a voltage larger than the oscilloscope can handle. Its high input resistance is also useful because we do not want to draw significant current from the circuit that we are studying since this would change the operation of the circuit. When one uses such an attenuator, the actual circuit is as drawn in Figure I-19b, in which C is the capacitance measured between the terminals of the oscilloscope, that is, the input capacitance of the oscilloscope plus the leads.

Let us solve for the output voltage of the realistic circuit, using Kirchhoff's laws.

This is the first circuit that we have considered in which there are several current paths, so that the application of the first law is not trivial. We will set up the equations arising from the circuit in two ways; the first using *real* currents, the second using *loop* currents. Both techniques arrive at identical solutions, and the choice of which to use depends on which method appeals to the student in a given instance.

For *real* currents, one defines a current, including its direction, in every branch of the circuit. This assumes that one has first reduced all parallel or series passive elements to their simplest configuration (see any elementary book in circuits if this poses a problem to you). If the direction chosen for the current in a given branch is incorrect, the results will simply give a negative value to the current. Kirchhoff's first law is then applied at the nodes (junctions) of the circuit (Fig. 1-20). A current entering a node (according to your assignment) is positive; one leaving a node is negative:

$$i_1 - i_2 - i_3 = 0 \qquad \text{at node } X.$$

The same equation is obtained at node Y with the signs reversed; thus, no new information is gained. In general, if there are N nodes, there will be $N - 1$ equations arising from the application of the first law. Kirchhoff's second law is then applied around closed voltage loops. These loops need not follow current flow. If the direction around a voltage loop is such that

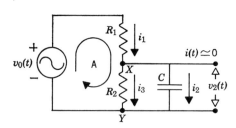

FIGURE I-20. Kirchhoff's laws applied to an uncompensated attenuator.

THE UNCOMPENSATED ATTENUATOR

one goes through an impedance in the same direction as the presumed current, the voltage drop is positive. If not, then it is negative. If one goes through a voltage source from negative to positive, the voltage gain is positive. If not, then it is negative. Therefore, for voltage loop A, we have

$$v_0(t) = i_1 R_1 + i_3 R_2.$$

A loop taken through R_2 and C in a clockwise direction gives

$$0 = -i_3 R_2 + q_2(t)/C.$$

Finally, a loop through the capacitor and the output voltage terminals in a counterclockwise direction gives

$$v_2(t) = q_2(t)/C.$$

We proceed by eliminating i_1 and i_3, since we are interested in the output voltage, which can be obtained from i_2 through

$$i_2(t) = \frac{dq_2}{dt}.$$

So,

$$v_0(t) = (i_3 + i_2)R_1 + i_3 R_2,$$

$$i_3(t) = \frac{q_2(t)}{R_2 C},$$

and

$$\frac{v_0(t)}{R_1} = \frac{(R_1 + R_2)}{R_1 R_2 C} q_2(t) + \frac{dq_2}{dt}.$$

The same results can be obtained, perhaps more easily, through the use of *loop* currents. These are artificial currents defined for closed current paths of the circuit (Fig. 1-21). The current in a given element is the algebraic sum of the loop currents in that element (see the Appendix to this chapter

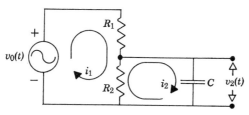

FIGURE 1-21. Solution of an uncompensated attenuator through the use of loop currents.

PASSIVE ELEMENTS: DIFFERENTIAL EQUATIONS

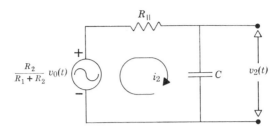

FIGURE I-22. Equivalent circuit for an uncompensated attenuator.

for a statement of the Superposition theorem). Now apply Kirchhoff's second law, for convenience following the current loops:

$$v_0(t) = i_1 R_1 + i_1 R_2 - i_2 R_2;$$
$$0 = i_2 R_2 - i_1 R_1 + q_2(t)/C.$$

The output voltage is

$$v_2(t) = q_2(t)/C.$$

We eliminate i_1 to obtain the differential equation for $q_2(t)$:

$$i_1 = q_2/R_2 C + i_2;$$
$$v_0(t) + i_2 R_2 - (R_1 + R_2)(q_2/R_2 C + i_2) = 0;$$
$$\frac{dq_2}{dt} + \frac{(R_1 + R_2)}{R_1 R_2 C} q_2 = \frac{v_0(t)}{R_1}.$$

This is the same result that was obtained by the method of real currents. Let us define $1/R_\| = 1/R_1 + 1/R_2$ and $\tau_\| = R_\| C$. We have, then,

$$\frac{dq_2}{dt} + \frac{q_2}{\tau_\|} = \frac{v_0(t)}{R_1}.$$

This is the type of equation we had before and could be solved in the same way.

We note here that we could replace the above circuit by the *equivalent circuit* of Fig. 1-22. This is possible since

$$\frac{dq_2}{dt} + \frac{q_2}{R_\| C} = \frac{v_0(t)}{R_1} = \frac{1}{R_\|} \left(\frac{R_\|}{R_1} v_0(t) \right)$$

and

$$\frac{R_\|}{R_1} = \left(\frac{R_1 R_2}{R_1 + R_2} \right) \frac{1}{R_1} = \frac{R_2}{R_1 + R_2}.$$

THE UNCOMPENSATED ATTENUATOR

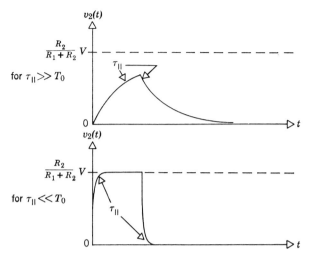

FIGURE 1-23. Response of an uncompensated attenuator to an input rectangular pulse, for $\tau_{\parallel} \gg T_0$ and $\tau_{\parallel} \ll T_0$.

The response to a square-wave pulse is found from the equivalent circuit to be

$$v_2(t) = \frac{R_2}{R_1 + R_2} V(1 - e^{-t/\tau_{\parallel}}) \quad \text{for} \quad 0 \leq t \leq T_0$$

and

$$v_2(t) = \frac{R_2}{R_1 + R_2} V(1 - e^{-T_0/\tau_{\parallel}}) e^{-(t-T_0)/\tau_{\parallel}} \quad \text{for} \quad T_0 \leq t.$$

We see in Fig. 1-23 that the probe-plus-oscilloscope circuit may distort the input pulse considerably.

The response of this circuit to a sinusoidal input $v_0(t) = V \sin(\omega t + \phi)$ may also be found from the equivalent circuit. The steady-state response is the one of interest and is found to be

$$v(t)_{\text{s.t.}} = \frac{R_2}{R_1 + R_2} \frac{V}{\sqrt{1 + (\omega \tau_{\parallel})^2}} \sin(\omega t + \phi - \tan^{-1} \omega \tau_{\parallel}).$$

(decrease in amplitude) (lag in phase)

It is possible, however, to eliminate the above distortion effects by placing an appropriate capacitor across R_1.

21

PASSIVE ELEMENTS: DIFFERENTIAL EQUATIONS

THE COMPENSATED ATTENUATOR

Let us consider the circuit in Fig. 1-24. Kirchhoff's second law gives us

1. $\quad v_0(t) - i_1(R_1 + R_2) + i_3 R_1 + i_2 R_2 = 0.$

2. $\quad i_1 R_2 - i_2 R_2 - \dfrac{q_2}{C_2} = 0.$

3. $\quad i_1 R_1 - i_3 R_1 - \dfrac{q_3}{C_1} = 0.$

The output voltage is given by

$$v_2(t) = \frac{q_2(t)}{C_2}.$$

Hence, we must solve for q_2.

We could also write Kirchoff's second law for the outer loop (or combine the above three equations to obtain it). This equation is better than (3) since (3) does not involve q_2 or i_2:

$$v_0(t) - \frac{q_3}{C_1} - \frac{q_2}{C_2} = 0.$$

Differentiating, we have

$$\frac{dv_0(t)}{dt} - \frac{i_3}{C_2} - \frac{i_2}{C_2} = 0, \quad \text{or} \quad i_3 = C_1 \frac{dv(t)}{dt} - \frac{C_1}{C_2} i_2.$$

From equation (2) we have

$$i_1 = i_2 + \frac{q_2}{R_2 C_2}.$$

Substituting these two equations into equation (1) gives

$$v_0(t) - (R_1 + R_2)\left(i_2 + \frac{q_2}{R_2 C_2}\right) + R_1\left(C_1 \frac{dv_0}{dt} - \frac{C_1}{C_2} i_2\right) + i_2 R_2 = 0;$$

$$i_2 \left[R_1 \left(\frac{C_1 + C_2}{C_2}\right)\right] + \left(\frac{R_1 + R_2}{R_2 C_2}\right) q_2 = v_0(t) + R_1 C_1 \frac{dv}{dt},$$

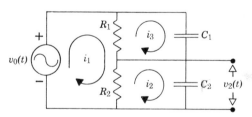

FIGURE 1-24. Compensated attenuator.

THE COMPENSATED ATTENUATOR

or
$$\frac{dq_2}{dt} + \frac{q_2}{\tau_{\|}} = C_S\left(\frac{v_0}{\tau_1} + \frac{dv_0}{dt}\right),$$

where
$$\tau_{\|} = R_{\|}C_{\|},$$
$$\frac{1}{R_{\|}} = \frac{1}{R_1} + \frac{1}{R_2},$$
$$C_{\|} = C_1 + C_2,$$
$$\frac{1}{C_S} = \frac{1}{C_1} + \frac{1}{C_2},$$

and
$$\tau_1 = R_1 C_1.$$

What is the response of this circuit to a step function? Obviously, the differential equation would present difficulties with the dv_0/dt in the inhomogeneous term since this would be infinite when the step is turned on. However, we can get around this by realizing that the step function is an idealization, an approximation to reality. Actually, there will always be some capacitance and resistance so that the exponential rise is more realistic:

$$v_0(t) = V(1 - e^{-t/\tau_R}).$$

The step function may be taken as a limiting case of this:

$$Vu(t) = \lim_{\tau_R \to 0} V(1 - e^{-t/\tau_R}).$$

We may solve the problem first with the exponential rise and then take the limit in the solution to obtain the result for the step. The differential equation becomes

$$\frac{dq_2}{dt} + \frac{q_2}{\tau_{\|}} = C_S \frac{V}{\tau_1}(1 - e^{-t/\tau_R}) + \frac{C_S}{\tau_R} V e^{-t/\tau_R}.$$

The homogeneous solution is

$$q_h = A e^{-t/\tau_{\|}}$$

The trial function for the particular solution is taken to be

$$q_p = B e^{-t/\tau_R} + D,$$
$$\frac{dq_p}{dt} = -\frac{B}{\tau_R} e^{-t/\tau_R}.$$

23

PASSIVE ELEMENTS: DIFFERENTIAL EQUATIONS

Substitution into the differential equation gives

$$-\frac{B}{\tau_R}e^{-t/\tau_R} + \frac{Be^{-t/\tau_R}}{\tau_\parallel} + \frac{D}{\tau_\parallel} = \frac{C_S V}{\tau_1}(1 - e^{-t/\tau_R}) + \frac{C_S}{\tau_R}Ve^{-t/\tau_R},$$

or

$$B\left(\frac{1}{\tau_\parallel} - \frac{1}{\tau_R}\right)e^{-t/\tau_R} + \frac{D}{\tau_\parallel} = C_S V\left(\frac{1}{\tau_R} - \frac{1}{\tau_1}\right)e^{-t/\tau_R} + \frac{C_S V}{\tau_1}.$$

If this equation is to be valid for *all* times, then we must have

$$D = \frac{C_S V \tau_\parallel}{\tau_1};$$

$$B = \frac{C_S V \tau_\parallel}{\tau_1}\left(\frac{\tau_1 - \tau_R}{\tau_R - \tau_\parallel}\right).$$

Let us use initial conditions

$$q_2 = 0 \qquad \text{at } t = 0.$$

Then

$$q_2(t) = Ae^{-t/\tau_\parallel} + \frac{C_S V \tau_\parallel}{\tau_1}\left(\frac{\tau_1 - \tau_R}{\tau_R - \tau_\parallel}\right)e^{-t/\tau_R} + \frac{C_S V \tau_\parallel}{\tau_1};$$

$$A = \left[\left(\frac{\tau_R - \tau_1}{\tau_R - \tau_\parallel}\right) - 1\right]\frac{C_S V \tau_\parallel}{\tau_1} = \left[\frac{\tau_\parallel - \tau_1}{\tau_R - \tau_\parallel}\right]\frac{C_S V \tau_\parallel}{\tau_1}.$$

The final solution is

$$q_2(t) = \frac{C_S V \tau_\parallel}{\tau_1}\left[1 - \left(\frac{\tau_\parallel - \tau_1}{\tau_\parallel - \tau_R}\right)e^{-t/\tau_\parallel} - \left(\frac{\tau_1 - \tau_R}{\tau_\parallel - \tau_R}\right)e^{-t/\tau_R}\right]$$

and

$$v_2(t) = \frac{q_2}{C_2}.$$

Taking the limit as $\tau_R \to 0$ yields

$$v_2(t) = \frac{R_2}{R_1 + R_2}V\left[1 - \left(1 - \frac{\tau_1}{\tau_\parallel}\right)e^{-t/\tau_\parallel}\right]$$

since

$$\frac{C_S \tau_\parallel}{C_2 \tau_1}V = \frac{1}{C_2}\left(\frac{C_1 C_2}{C_1 + C_2}\right)\left[\left(\frac{R_1 R_2}{R_1 + R_2}\right)(C_1 + C_2)\right]\left(\frac{1}{R_1 C_1}\right)V = \frac{R_2}{R_1 + R_2}V.$$

Thus, if $1 - \tau_1/\tau_\parallel = 0$, the *time dependence disappears*. The *compensation condition* is then $\tau_1 = \tau_\parallel$. A little algebra shows that this condition leads also to $\tau_1 = \tau_2$.

THE COMPENSATED ATTENUATOR

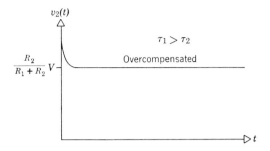

FIGURE 1-25. Overcompensation in a "compensated attenuator".

If $\tau_1/\tau_\| > 1$, then $[1 - (\tau_1/\tau_\|)]$ is negative, and we have the overcompensated curve shown in Fig. 1-25. If $\tau_1/\tau_\| < 1$, then $[1 - (\tau_1/\tau_\|)]$ is positive, and we have the undercompensated curve shown in Fig. 1-26.

We may solve for the response of the circuit to a sinusoidal input in the same way as before. With the input $V(t) = V \sin wt$ and the initial condition $q_2(0) = 0$, we find for the steady-state solution

$$v_2(t) = \frac{C_1}{C_1 + C_2} V \sqrt{\beta^2 + (1-\beta)^2 \omega^2 \tau_\|^2} \sin\left\{\omega t + \tan^{-1}\left[\frac{\omega \tau_\|(1-\beta)}{\beta}\right]\right\},$$

where

$$\beta = \frac{\dfrac{\tau_\|}{\tau_1} + \omega^2 \tau_\|^2}{1 + \omega^2 \tau_\|^2}.$$

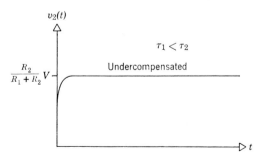

FIGURE 1-26. Undercompensation in a "compensated attenuator".

PASSIVE ELEMENTS: DIFFERENTIAL EQUATIONS

If $\tau_\perp = \tau_\parallel$, we have

$$\frac{C_1}{C_1 + C_2} = \frac{R_2}{R_1 + R_2}$$

and

$$v_2(t) = \frac{R_2}{R_1 + R_2} V \sin \omega t.$$

Because this is the compensation condition for the square wave and for a sinusoid input of arbitrary frequency ω, we know that it works for most functions since we may Fourier analyze any given function into a sum or integral of sinusoidal waves.*

The above analysis indicates the necessity of checking and compensating an oscilloscope probe before attempting to make any precise measurements. We will also find compensated attenuators useful in coupling various elements of a circuit together without introducing frequency dependence. (As an example, see the transistor binary, Chapter 11.)

RCL CIRCUITS

We now wish to consider circuits containing both capacitance and inductance. A new phenomenon is possible since we have a storage device for potential energy (capacitance) and a storage device for kinetic energy (inductance). Energy may be transferred back and forth between these two, giving rise to a resonant condition at some frequency. Let us consider a series circuit (Fig. 1-27).

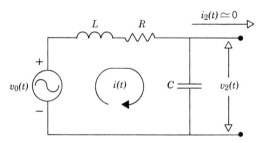

FIGURE 1-27. Series RCL circuit.

* For example, see I. S. Sokolnikoff and R. M. Redheffer, *Mathematics of Physics and Modern Engineering*, McGraw-Hill, New York (1966), Chapter 1.

RCL CIRCUITS

Kirchhoff's second law provides us with the equations

$$v_0(t) - L\frac{di}{dt} - iR - \frac{q}{C} = 0;$$

$$v_2(t) = \frac{q(t)}{C}.$$

In terms of the charge q we have an inhomogeneous, second-order, linear differential equation with constant coefficients:

$$\frac{d^2q}{dt^2} + \frac{R}{L}\frac{dq}{dt} + \frac{q}{LC} = \frac{v_0(t)}{L}.$$

For the homogeneous solution, we try $q_h(t) = Ae^{at}$, which gives us the secular equation

$$a^2 + \frac{R}{L}a + \frac{1}{LC} = 0,$$

or

$$a = -\gamma \pm j\omega_1,$$

where

$$\omega_1 = \sqrt{\omega_0^2 - \gamma^2};$$

$$\omega_0 = \frac{1}{\sqrt{LC}}, \quad \text{and} \quad \gamma = \frac{R}{2L}.$$

Let us consider the response of the circuit to a step input for the case $\omega_0 > \gamma$. The particular solution is $q_p(t) = \text{constant} = CV$. Hence,*

$$q(t) = e^{-\gamma t}(A \sin \omega_1 t + D \cos \omega_1 t) + CV.$$

Let

$$q(0) = 0 \quad \text{and} \quad i(0) = 0.$$

Then

$$q(t) = CV\left[1 - \left(\frac{\gamma}{\omega_1}\sin \omega_1 t + \cos \omega_1 t\right)e^{-\gamma t}\right],$$

or

$$q(t) = CV\left[1 - \frac{\omega_0}{\omega_1}e^{-\gamma t}\sin\left(\omega_1 t + \tan^{-1}\frac{\omega_1}{\gamma}\right)\right].$$

The output voltage is then

$$v_2(t) = V\left\{1 - \left[\frac{\omega_0}{\omega_1}\sin\left(\omega_1 t + \tan^{-1}\frac{\omega_1}{\gamma}\right)\right]e^{-\gamma t}\right\}.$$

* Recall that $\cos \theta = (e^{j\theta} + e^{-j\theta})/2$ and $\sin \theta = (e^{j\theta} - e^{-j\theta})/2j$.

PASSIVE ELEMENTS: DIFFERENTIAL EQUATIONS

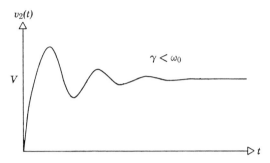

FIGURE 1-28. Output of a series RCL circuit for an input step function. The output is taken across the capacitor and the circuit is under damped.

The response in this case is similar to that of the integrating circuit but there is now an oscillation (or "ringing") of the output at frequency ω_1 which damps out with the time constant $1/\gamma$ (Fig. 1-28).

We note that for zero resistance the oscillations are undamped and we have an oscillating *tank* circuit

$$v_2(t) = V[1 - \cos \omega_0 t]$$

with resonant frequency $\omega_0 = 1/\sqrt{LC}$.

A parameter which indicates how close a circuit is to a resonant circuit is the Q of a circuit, where

$$Q = \frac{\omega_0}{2\gamma} = \frac{\omega_0 L}{R} \text{ (series } RCL \text{ circuit)}.$$

Then the oscillation frequency may be written

$$\omega_1 = \omega_0 \sqrt{1 - \left(\frac{1}{2Q}\right)^2}.$$

The resistance in the circuit dissipates the oscillatory energy and finally damps out the ringing.

If the capacitor and resistor are interchanged, the output voltage (Fig. 1-29) is

$$v_2(t) = \frac{RV}{\omega_1 L} e^{-\gamma t} \sin \omega_1 t.$$

A practical measure of the decay is the *logarithmic decrement D:*

$$D = \ln \frac{v_2(t)}{v_2(t+T)} = \gamma T,$$

where $T = 1/f = 2\pi/\omega_1$ is the period of oscillation.

RCL CIRCUITS

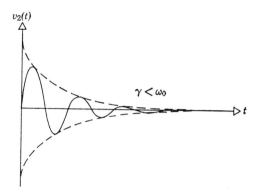

FIGURE 1-29. Output of a series *RCL* circuit for an input step function. The output is taken across the resistor and the circuit is under damped.

There are two other cases (different values of *R*, *L* and *C*) possible for this circuit. They are $\omega_0 < \gamma$ and $\omega_0 = \gamma$.

In the first case we have $\omega_1 = j\beta$, where $\beta = \sqrt{\gamma^2 - \omega_0^2}$, giving us the results shown in Fig. 1-30 for the output across the resistance:

$$v_2(t) = \frac{RV}{\beta L} e^{-\gamma t} \sinh \beta t.$$

Critical damping occurs when $\omega_0 = \gamma$ or $\omega_1 = 0$. It is evident we cannot use the above solutions since they would result in an infinite output. We must solve the differential equation again:

$$\frac{d^2 q}{dt^2} + 2\omega_0 \frac{dq}{dt} + \omega_0^2 q = \frac{v_0(t)}{L}.$$

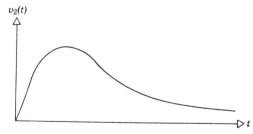

FIGURE 1-30. Output of a series *RCL* circuit for an input step function. The output is taken across the resistor and the circuit is over damped.

29

PASSIVE ELEMENTS: DIFFERENTIAL EQUATIONS

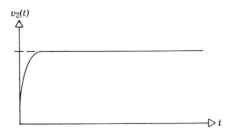

FIGURE 1-31. Output of a series RCL circuit for an input step function. The output is taken across the capacitor and the circuit is critically damped.

The secular equation for the homogeneous equation now has a double root:

$$q_h(t) = Ae^{at};$$
$$a^2 + 2\omega_0 a + \omega_0^2 = 0;$$
$$(a + \omega_0)^2 = 0; \ a = -\omega_0 = -\gamma.$$

In this case we have only one arbitrary constant. However, the following is also a solution:

$$q_h(t) = Bte^{at}.$$

Hence, for the step input,

$$q(t) = (A + Bt)e^{-\gamma t} + CV.$$

The initial conditions $[q(0) = 0, i(0) = 0]$ give the curve shown in Fig. 1-31 for the voltage $v_2(t)$ across the capacitor:

$$v_2(t) = V[1 - (1 + \gamma t)e^{-\gamma t}].$$

PHASORS AND COMPLEX IMPEDANCE

We now illustrate a method used extensively by engineers to find the steady-state response of the series RCL circuit to a sinusoidal input. The differential equation which must be solved is

$$L\frac{dq^2}{dt^2} + R\frac{dq}{dt} + \frac{q}{C} = V\sin(\omega t + \phi).$$

We may simplify the analysis by using complex variables. The important point to note is that, since the real and imaginary parts of a complex number

PHASORS AND COMPLEX IMPEDANCE

are independent, a necessary and sufficient condition for the complex differential equation

$$\frac{d^2z}{dt^2} + A\frac{dz}{dt} + Bz = w$$

where

$$z(t) = x(t) + jy(t)$$
$$w(t) = u(t) + jv(t)$$

to be valid is that the following equations be satisfied:

$$\frac{d^2x}{dt^2} + A\frac{dx}{dt} + Bx = u$$

and

$$\frac{d^2y}{dt^2} + A\frac{dy}{dt} + By = v.$$

Thus, we may solve the complex problem and then take the real or imaginary result for the solution of our original problem.

Hence, let us use

$$\hat{v}_0(t) = Ve^{j(\omega t + \phi)},$$

which we call a "phasor," and find a complex solution for $\hat{q}(t)$. Then $q(t) = \text{Im } \hat{q}(t)$ since $v_0(t) = \text{Im } \hat{v}_0(t)$. (Recall that $e^{j\theta} = \cos\theta + j\sin\theta$.)

The equation to be solved is

$$L\frac{d^2\hat{q}}{dt^2} + R\frac{d\hat{q}}{dt} + \frac{\hat{q}}{C} = \hat{v}_0.$$

We try a solution

$$\hat{q}(t) = Ae^{j(\omega t + \phi)},$$

where A is a complex number. Substituting this into the differential equation gives us

$$A\left(-\omega^2 L + j\omega R + \frac{1}{C}\right) = V.$$

Thus,

$$A = \frac{V}{j\omega\left[R + j\left(\omega L - \frac{1}{\omega C}\right)\right]}.$$

The complex charge (phasor) is then

$$\hat{q}(t) = \frac{Ve^{j(\omega t + \phi)}}{j\omega[R + jX]},$$

where

$$X = X_L - X_C$$

31

PASSIVE ELEMENTS: DIFFERENTIAL EQUATIONS

is the *reactance* of the circuit,
$$X_L = \omega L$$
is the *inductive reactance*, and
$$X_C = \frac{1}{\omega C}$$
is the *capacitive reactance*. The complex current (phasor) is
$$\hat{\imath}(t) = \frac{V e^{j(\omega t + \phi)}}{R + jX}.$$
We may write this in the following way:
$$\hat{V} = \hat{\imath}\hat{Z},$$
where
$$\hat{Z} = R + jX$$
is called the *complex impedance*. The above equation is sometimes called Ohm's law for ac circuits. We may also write the complex impedance in terms of its magnitude and phase:
$$Z = \sqrt{R^2 + X^2}\, e^{j(\tan^{-1}(X/R))}.$$
The voltages across the various elements are now found directly:
$$v_R(t) = \mathrm{Im}\,(R\hat{\imath}) = \mathrm{Im}\,\frac{RV}{\sqrt{R^2 + X^2}}\, e^{j(\omega t + \phi - \tan^{-1}(X/R))},$$
or
$$v_R(t) = \frac{RV}{\sqrt{R^2 + X^2}} \sin\!\left(\omega t + \phi - \tan^{-1}\frac{X}{R}\right).$$
We note that the voltage $v_R(t)$ is in phase with the current $i(t)$:
$$i(t) = \mathrm{Im}\,\hat{\imath} = \frac{V}{\sqrt{R^2 + X^2}} \sin\!\left(\omega t + \phi - \tan^{-1}\frac{X}{R}\right).$$
The voltage across the capacitor is
$$v_C(t) = \mathrm{Im}\,\frac{\hat{q}}{C} = \mathrm{Im}\left[\frac{\hat{\imath}}{j\omega C}\right],$$
or
$$v_C(t) = \frac{X_C V}{\sqrt{R^2 + X^2}} \sin\!\left(\omega t + \phi - \tan^{-1}\frac{X}{R} - \frac{\pi}{2}\right).$$

PHASORS AND COMPLEX IMPEDANCE

Here the voltage across the capacitor lags the current through it by 90°. The voltage across the inductance is

$$v_L(t) = \text{Im}\left(L\frac{di}{dt}\right) = \text{Im}(j\omega L\hat{i}),$$

or

$$v_L(t) = \frac{X_L V}{\sqrt{R^2 + X^2}} \sin\left(\omega t + \phi - \tan^{-1}\frac{X}{R} + \frac{\pi}{2}\right).$$

The voltage across the inductor leads the current through it by 90°. In Navy electronics the phases are remembered by the phrase

"ELI the ICEman."

To see the resonant phenomenon more clearly, we consider the power absorbed by the resistor. This will be proportional to the voltage squared, or

$$P \propto \frac{R^2}{R^2 + X^2}.$$

At resonance ($\omega = \omega_0 = 1/\sqrt{LC}$), the reactance is

$$X = \omega L - \frac{1}{\omega C} = \omega L\left(1 - \frac{1}{\omega^2 LC}\right) = 0.$$

Thus, the power absorbed is a maximum at resonance.

Another approach is to consider the *complex admittance*

$$\hat{Y} = \frac{1}{\hat{Z}} = g + jb,$$

where

$$g(\omega) = \frac{R}{R^2 + X^2}$$

is the *conductance* and

$$b(\omega) = -\frac{X}{R^2 + X^2}$$

is the *susceptance* (Fig. 1-32).

We may determine the Q of the circuit in the following way:

$$X = \omega L\left(\frac{\omega^2 - \omega_0^2}{\omega^2}\right) = \omega L \frac{(\omega + \omega_0)(\omega - \omega_0)}{\omega \omega}.$$

Near $\omega = \omega_0$ we may write

$$X \simeq \omega_0 L\left[\frac{2(\omega - \omega_0)}{\omega_0}\right].$$

33

PASSIVE ELEMENTS: DIFFERENTIAL EQUATIONS

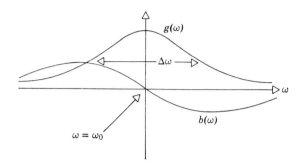

FIGURE 1-32. Relationship between conductance and susceptance.

At the half-power points, $X^2 = R^2$. Hence, $X = \pm R$ and

$$\frac{X}{\omega_0 L} = \pm \frac{R}{\omega_0 L} = \pm \frac{1}{Q} = \frac{2(\omega - \omega_0)}{\omega_0}.$$

Calling $2|\omega - \omega_0| = \Delta\omega$, we have $|\Delta\omega|/\omega_0 = 1/Q$, or

$$\frac{|\Delta f|}{f_0} = \frac{1}{Q}.$$

A high Q circuit has a high *selectivity* with respect to frequency.

There is one caution to be observed when using the complex phasor method. If the power is desired, we must calculate

$$P = iv,$$

or

$$P = (\text{Im } \hat{\imath})(\text{Im } \hat{v}).$$

We might be tempted to write

$$P = \text{Im } (\hat{\imath}\hat{v}),$$

but this is incorrect since

$$\hat{P} = \hat{\imath}\hat{v} = (\text{Re } \hat{\imath} + j \text{ Im } \hat{\imath})(\text{Re } \hat{v} + j \text{ Im } \hat{v})$$

$$= (\text{Re } \hat{\imath})(\text{Re } \hat{v}) - (\text{Im } \hat{\imath} \text{ Im } \hat{v}) + j[(\text{Re } \hat{\imath})(\text{Im } \hat{v}) + (\text{Im } \hat{\imath})(\text{Re } \hat{v})].$$

Hence,

$$\text{Im } (\hat{\imath}\hat{v}) = (\text{Re } \hat{\imath})(\text{Im } \hat{v}) + (\text{Im } \hat{\imath})(\text{Re } \hat{v}) \neq (\text{Im } \hat{\imath})(\text{Im } \hat{v}).$$

(A meaningful result may, however, be obtained for the time average of the complex power. See Problem 1-13.)

Q OF A CIRCUIT

Q OF A CIRCUIT

The concept of the Q of a circuit may be given a very physical interpretation which may be extended to cover various resonant devices, such as resonant microwave cavities. The more general definition is as follows:

$$Q = 2\pi \left(\frac{\text{average energy stored per cycle}}{\text{energy loss per cycle}} \right).$$

We now illustrate the applicability of this definition to circuits by calculating the Q of the series LRC circuit.

The average energy stored per cycle in the capacitor is

$$W_c = \frac{1}{2} \frac{q^2}{C}.$$

The charge was found to be

$$q(t) = \frac{V}{\omega\sqrt{R^2 + X^2}} \sin\left(\omega t + \phi - \tan^{-1}\frac{X}{R} - \frac{\pi}{2}\right).$$

Hence, the average energy stored per cycle is

$$W_c = \frac{1}{T} \int_0^T \frac{q^2}{2C} dt,$$

or

$$W_c = \frac{V^2}{2\omega^3 C(R^2 + X^2)T} \int_{\phi - (\tan^{-1}(X/R))}^{2\pi + \phi - \tan^{-1}(X/R)} \cos^2 u \, du$$

but

$$\int_a^{a+2\pi} \cos^2 u \, du = \pi.$$

Hence,

$$W_c = \frac{V^2}{4\omega^2 C(R^2 + X^2)}.$$

For the inductor, we have

$$W_L = \tfrac{1}{2} L i^2;$$

$$i(t) = \frac{V}{\sqrt{R^2 + X^2}} \sin\left(\omega t + \phi - \tan^{-1}\frac{X}{R}\right);$$

$$W_L = \frac{1}{T} \int_0^T \tfrac{1}{2} L i^2 \, dt,$$

35

PASSIVE ELEMENTS: DIFFERENTIAL EQUATIONS

or

$$W_L = \frac{LV^2}{2\omega(R^2 + X^2)T} \int_{\phi-\tan^{-1}(X/R)}^{2\pi+\phi-\tan^{-1}(X/R)} \sin^2 u \, du;$$

but

$$\int_a^{a+2\pi} \sin^2 u \, du = \pi.$$

Thus, the average energy stored per cycle in the inductor is

$$W_L = \frac{LV^2}{4(R^2 + X^2)}.$$

The energy dissipated per cycle in the resistor is given by

$$W_R = \int_0^T Ri^2 \, dt,$$

or

$$W_R = \frac{RV^2 T}{2(R^2 + X^2)}.$$

The Q of the circuit as given by the general definition is

$$Q = 2\pi \frac{\dfrac{V^2}{4\omega^2 C(R^2 + X^2)} + \dfrac{LV^2}{4(R^2 + X^2)}}{\dfrac{R}{2} \dfrac{V^2}{(R^2 + X^2)} \dfrac{2\pi}{\omega}},$$

or

$$Q = \frac{1}{2R}\left[\omega L + \frac{1}{\omega C}\right].$$

It can be seen that this definition of the Q of the circuit leads to a frequency-dependent Q. However, in the vicinity of $\omega = \omega_0$, we may approximate the expression by

$$Q = \left(\frac{L}{R}\right)\frac{1}{2}\left[\omega + \frac{\omega_0^2}{\omega}\right] \simeq \frac{\omega_0 L}{R}.$$

This result is identical to our first definition of the Q of the series LRC circuit.

APPENDIX

Several theorems can be derived from Kirchhoff's laws as applied to linear circuits, theorems that can greatly simplify the analysis of linear circuits. The most important of these theorems are stated here:

APPENDIX

(1) The Superposition Theorem

In a circuit containing several independent sources, the effect of these sources on a passive element of the circuit can be obtained by algebraic superposition of the effect of each of the sources acting alone, with all other sources replaced by their internal impedances.

(2) Thevenin's Theorem

With respect to any two terminals of a linear circuit, the circuit can be replaced by an ideal voltage source in series with an impedance. (Fig. 1-33). The value of

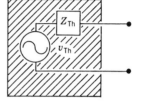

FIGURE 1-33. Equivalent circuit generated by Thevenin's theorem.

the voltage source is given by the open-circuit voltage appearing across the terminals, while the impedance is that appearing between the terminals.

(3) Norton's Theorem

With respect to any two terminals of a linear circuit, the circuit can be replaced by an ideal current source in parallel with an impedance (Fig. 1-34). The value of

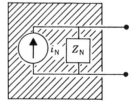

FIGURE 1-34. Equivalent circuit generated by Norton's theorem.

the current source is given by the current which flows between the short-circuited terminals, while the impedance is that which appears across the terminals.

(4) Short-circuit Theorem

Thevenin's theorem and Norton's theorem together give the short-circuit theorem, which states that if v_{open} is the open-circuit voltage that appears across a pair of terminals of a linear system, and i_{sc} is the current that flows between the terminals when they are short-circuited, then

$$v_{\text{open}} = i_{\text{sc}} Z = i_{\text{sc}}/Y,$$

PASSIVE ELEMENTS: DIFFERENTIAL EQUATIONS

where Z is the impedance between the terminals and Y is the admittance between the terminals.

As an example of the use of the above theorems, we can apply them to the solution of the following circuit:

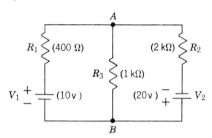

FIGURE 1-35. A parallel dc circuit.

This circuit can certainly be solved by direct application of Kirchhoff's laws, but by using the short-circuit theorem, the same result can be obtained more easily. Suppose we wish to find the voltage across R_3. Choose points A and B as the terminals. We can obtain i_{sc} by placing a short between A and B, and using the superposition theorem:

$$i_{sc} = V_1/R_1 + (-V_2)/R_2 = 15 \text{ ma}.$$

With the sources replaced by their internal impedances (assumed to be zero), the admittance between A and B is that of the three resistors in parallel, or

$$Y = 1/R_1 + 1/R_2 + 1/R_3 = \frac{1}{250 \text{ ohms}}.$$

Then

$$v_{open} = i_{sc}/Y = (15 \text{ ma})(250 \text{ ohms}) = 3.75 \text{ volts}.$$

PROBLEMS

1. Apply Kirchhoff's laws to the following dc circuit to find the voltage across resistor R_2.

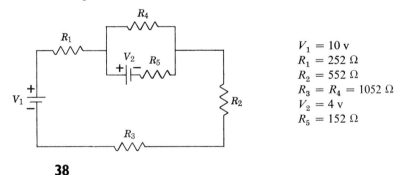

$V_1 = 10$ v
$R_1 = 252 \ \Omega$
$R_2 = 552 \ \Omega$
$R_3 = R_4 = 1052 \ \Omega$
$V_2 = 4$ v
$R_5 = 152 \ \Omega$

PROBLEMS

2. Solve the circuit appearing in Fig. 1-35 by direct application of Kirchhoff's laws.
3. Set up (but do not solve) the equations for the current flowing through resistor R_5 in the following bridge circuit. One way to do this is to put the equations obtained from Kirchhoff's laws in a determinant form.

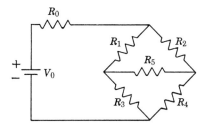

4. In a situation in which it is necessary to use a blocking capacitor to protect one part of a circuit from a dc voltage present in another part of the circuit, it is important to know how much distortion will occur as a signal is passed through the capacitor. Assume a rectangular pulse incident upon an RC blocking (differentiating) circuit, with width T_0 and voltage V. Derive a relationship that will give the time constant necessary to insure no more than a 5% "droop" to the output pulse. (The droop may be defined as

$$[v_2(t = 0) - v_2(t = T_0)]/v_2(t = 0).)$$

5. Design a differentiating circuit using resistors and inductors. Compare this circuit with an RC differentiating circuit.
6. Design an integrating circuit utilizing resistors and inductors. Compare this circuit with an RC integrating circuit.
7. Suppose that one had a "black box" in which there was a circuit which behaved like a differentiating circuit when fed a rectangular pulse. How would you go about telling whether it was an RC or RL differentiating circuit?
8. In an RC differentiating circuit, it is obvious that there can be no net flow of charge around the circuit. Verify this by showing by direct integration that the area under the pulse plus the area of the undershoot equals zero.
9. Find the output voltage as a function of time for a differentiating RC circuit when the input wave behaves in the following manner:

$$v(t) = 0 \qquad t < 0;$$
$$v(t) = v_0 t \qquad t > 0.$$

Find an approximate form for the difference between the input and the output for times $0 \leqslant t \ll \tau$. What is the limiting value of the output as t becomes very much larger than the time constant, τ?

10. What is the input impedance for an RC differentiating circuit?
11. What is the output impedance for an RC differentiating circuit?

PASSIVE ELEMENTS: DIFFERENTIAL EQUATIONS

12. Find the input and output impedances of an *RC* integrating circuit.
13. Obtain the time average of a complex power. How does this compare with the usual definition of power?

Experiment 1

RC and *RL* Circuits

The first experiment will study the integrating and differentiating circuits basic to more complicated passive networks, and will check the results with those predicted by the analysis. The concept of compensation will be studied through the important compensated attenuator. It is expected that the results noted in the experiments may not be exactly those expected, and the analysis of the deviations from theory must enter prominently into your discussion of the results.

Equipment:
1. Oscilloscope
2. Sine- and square-wave generator
3. 2 Decade resistor boxes, 1 ohm to 100,000 ohms
4. 2 Decade capacitor boxes, 100 pf to 0.01 μf
5. 1 Decade inductor, or a fixed inductor with $L = 0.01$ h

I. Precaution
   Be sure that the oscilloscope is grounded to the wave generator.

II. Square-wave input

   A. High-pass, or differentiating, circuit

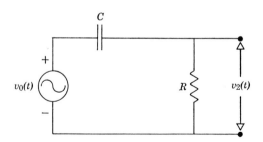

   1. Set the generator to square wave, at a frequency of 10 kHz.
   2. Observe and sketch the output wave form for the following time constants. Also, measure significant amplitudes and times and put these on your diagrams.

EXPERIMENT I

a. Use $\tau = RC = 1000\ \mu\text{sec}$
   $R = 10\ \text{k}\Omega$
   $C = 0.1\ \mu\text{f}$
b. Use $\tau = RC = 10\ \mu\text{sec}$
   $R = 5\ \text{k}\Omega$
   $C = 0.002\ \mu\text{f}$
c. Use $\tau = RC = 0.1\ \mu\text{sec}$
   $R = 500\ \Omega$
   $C = 0.0002\ \mu\text{f}$

B. Low-Pass, or integrating, circuit

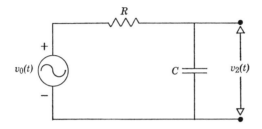

1. Same general instructions as in A
2. Use the following time constants.

   a. $\tau = RC = 0.1\ \mu\text{sec}$
      $R = 500\ \Omega$
      $C = 0.0002\ \mu\text{f}$
   b. $\tau = RC = 10\ \mu\text{sec}$
      $R = 5\ \text{k}\Omega$
      $C = 0.002\ \mu\text{f}$
   c. $\tau = RC = 35\ \mu\text{sec}$
      $R = 5\ \text{k}\Omega$
      $C = 0.007\ \mu\text{f}$

3. For $\tau = 35\ \mu\text{sec}$, measure the time required for the pulse to reach 0.63 of its final value. Also, measure the time (t_r) from the 10% to 90% points on the waveform.

   a. Use $t_r = 2.2\ RC$ to find the time constants of the circuit.
   b. Verify that $RC =$ time required to reach 0.63 of final value.

PASSIVE ELEMENTS: DIFFERENTIAL EQUATIONS

C. RL integrating circuit

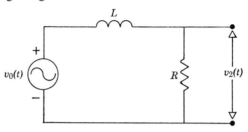

1. Use $R = 500 \; \Omega$; $L = 0.01 \; h$.
2. Observe and sketch waveform.

D. RL differentiating circuit

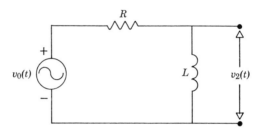

1. Use $R = 2 \; k\Omega$; $L = 0.01 \; h$.
2. Observe and sketch waveform. Compare with your calculations of the expected result.

E. Compensated attenuator
   1. Put together a compensated attenuator as shown schematically below, using the decade boxes provided.

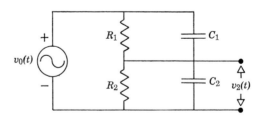

   2. Use the following values
   $R_1 = 9 \; k\Omega$ $C_1 = 0.001 \; \mu f$
   $R_2 = 1 \; k\Omega$ $C_2 = 0.009 \; \mu f$

EXPERIMENT I

    3. Observe waveform at v_2.
        a. Note attenuation ($v_0 = 10v_2$).
        b. Note lack of distortion when compensation condition is satisfied ($R_1C_1 = R_2C_2$).
        c. Vary C_1 both ways, noting distortion produced.
        d. Sketch results.

III. Sinusoidal input

    A. High-pass circuit (shown above)
        1. Using $R = 5$ kΩ, $C = 0.002$ μf, calculate $\omega\tau = 1$ to find half-power point.
        2. Measure input and output amplitude at several points near the frequency calculated in (1).
        3. Plot v_2/v_0 as a function of frequency.

    B. Low-pass circuit (shown above)
       Use same values as in A; follow same procedure.

ANALYSIS OF PASSIVE ELEMENTS LAPLACE TRANSFORMS

THE LAPLACE TRANSFORM*

As has been seen in the first part of the course, the solution of even fairly simple circuits may be quite complicated since they may consist of coupled integro-differential equations. For example, application of Kirchhoff's laws to the circuit in Fig. 2-1 leads to these equations:

$$v_0(t) - L\frac{di_1}{dt} - i_1 R + i_2 R = 0;$$

$$i_1 R - i_2 R - \frac{1}{C}\int i_2\,dt = 0;$$

$$v_2(t) = \frac{1}{C}\int i_2\,dt.$$

---

* See W. T. Thomson, *Laplace Transformation*, 2nd ed., Prentice-Hall, Inc., Englewood Cliffs, N.J. (1960), or M. R. Spiegel, Schaum's Outline Series, *Theory and Problems of Laplace Transforms*, Schaum Publishing Co., New York (1965) for good discussions of the Laplace transform method. See also P. A. McCollum and B. F. Brown, *Laplace transform tables and theorems*, Holt, Rinehart and Winston, Inc., New York, N.Y. (1965).

PASSIVE ELEMENTS: LAPLACE TRANSFORMS

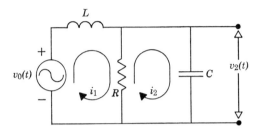

FIGURE 2.1. A series-parallel *RCL* circuit.

In the Laplace transform treatment, we make a transformation of variables from the real variable t (time) to a complex variable s (sometimes called the complex frequency).

The transformation is defined to be

$$\bar{F}(s) = \int_0^\infty f(t) e^{-st}\, dt$$

where $\bar{F}(s)$ is called the Laplace transform of $f(t)$. There are certain restrictions placed upon the function $f(t)$ in order for the integral to converge. In particular, $f(t)$ cannot grow more rapidly than e^{at}.

One of the most important functions to consider is the exponential:

$$f(t) = e^{-at}.$$

The transformation above gives us the Laplace transform of the exponential.

$$\bar{F}(s) = \int_0^\infty e^{-at} e^{-st}\, dt = \int_0^\infty e^{-(s+a)t}\, dt$$

$$= -\frac{e^{-(s+a)t}}{s+a} \Big|_0^\infty,$$

or

$$\bar{F}(s) = \frac{1}{s+a}.$$

We now wish to examine this function in the complex s plane. Consider the case in which a is a real number (Fig. 2-2).

A point in the complex plane where the function $\bar{F}(s)$ goes to infinity is called a *pole*. Looking at the inverse problem, if we have found the Laplace transform of a given output voltage to be of the above form, then we know that the output is a decaying exponential e^{-at}. Given an arbitrary Laplace

THE LAPLACE TRANSFORM

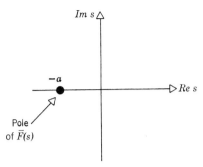

FIGURE 2.2. Pole-zero plot of a decaying exponential.

transform, one may determine the real function by means of the inverse transform. However, this involves a contour integral in the complex plane, which is beyond the level of this course. Happily, these inverse transforms have been worked out for all of the cases we shall be interested in, so we shall adopt the practice of just "looking up" the transform in a table.

Let us look again at the simple circuits: differentiator, integrator, compensated attenuator, and series RCL circuit. For the differentiator, we have the circuit shown in Fig. 2-3.

From Kirchhoff's laws we have the following equations to solve:

$$v_0 - \frac{1}{C} \int i \, dt - iR = 0;$$

$$v_2 = iR.$$

We transform these equations by multiplying by e^{-st} and integrating from 0 to ∞:

$$\int_0^\infty v_0 e^{-st} \, dt - \frac{1}{C} \left[\int_0^\infty \left(\int i \, dt \right) e^{-st} \, dt \right] - R \int_0^\infty i e^{-st} \, dt = 0;$$

$$\int_0^\infty v_2 e^{-st} \, dt = R \int_0^\infty i e^{-st} \, dt.$$

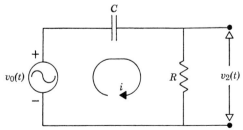

FIGURE 2-3. RC differentiating circuit.

47

PASSIVE ELEMENTS: LAPLACE TRANSFORMS

Or assuming the capacitor to be uncharged initially,

$$\bar{V}_0(s) - \frac{1}{C}\left(\frac{\bar{I}(s)}{s}\right) - \bar{I}(s)R = 0;$$

$$\bar{V}_2(s) = \bar{I}(s)R.$$

We have used the result of Problem 2-1 that

$$\mathscr{L}\left[\int_0^t f(t)\,dt\right] = \frac{\bar{F}(s)}{s}.$$

The transformed equations have the advantage of being algebraic, so we may easily solve them for the Laplace transform of the output voltage $\hat{V}_2(s)$. Thus,

$$\bar{V}_2(s) = \frac{sRC}{1+sRC}\bar{V}_0(s),$$

or

$$\bar{V}_2(s) = \bar{T}(s)\bar{V}_0(s),$$

where

$$\bar{T}(s) = \frac{s\tau}{1+s\tau} = \frac{s}{s+\dfrac{1}{\tau}}$$

is called the *transfer function* of the circuit. This function of the complex variable s has a *pole* at $s = -1/\tau$ and a *zero* at $s = 0$ (Fig. 2-4). We will find the poles and zeros of the transfer functions to be very useful in determining the character of the circuits. Notice the importance of the time constant.

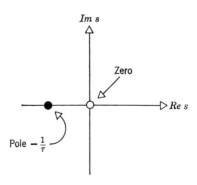

FIGURE 2-4. Pole-zero plot of the transfer function of an RC differentiating circuit.

THE LAPLACE TRANSFORM

Let us now find the response of the circuit to a step function. We must first find the Laplace transform of the unit step function:

$$u(t) = \begin{cases} 1 & 0 < t \\ 0 & t \leq 0. \end{cases}$$

Hence,

$$\bar{U}(s) = \int_0^\infty e^{-st}\,dt = -\frac{e^{-st}}{s}\bigg|_0^\infty = \frac{1}{s}.$$

The Laplace transform of the output voltage is

$$\bar{V}_2(s) = \frac{s}{s + \frac{1}{\tau}}\left(\frac{V}{s}\right) = \frac{V}{s + \frac{1}{\tau}};$$

thus,

$$v_2(t) = Ve^{-t/\tau},$$

which is the result found from the differential-equation approach.

If we wish the response of the circuit to a rectangular pulse, we have

$$v_0(t) = V[u(t) - u(t - T_0)].$$

To find the transform of $u(t - T_0)$, let us consider

$$\bar{F}(s) = \int_0^\infty e^{-st'}f(t')\,dt'.$$

Let $t' = t - T_0$; then

$$\bar{F}(s) = \int_{T_0}^\infty e^{-s(t-T_0)}f(t - T_0)\,dt$$

$$= e^{sT_0}\int_0^\infty e^{-st}f(t - T_0)u(t - T_0)\,dt;$$

$$e^{-sT_0}\bar{F}(s) = \int_0^\infty e^{-st}[f(t - T_0)u(t - T_0)]\,dt;$$

thus,

$$\mathscr{L}^{-1}[e^{-sT_0}\bar{F}(s)] = f(t - T_0)u(t - T_0).$$

This is called the *shifting theorem*. For $f(t - T_0) = u(t - T_0)$, we have

$$\mathscr{L}[u(t - T_0)] = \frac{e^{-sT_0}}{s}.$$

Thus, the Laplace transform of a rectangular pulse is

$$\bar{V}_0(s) = \frac{V(1 - e^{-sT_0})}{s}.$$

PASSIVE ELEMENTS: LAPLACE TRANSFORMS

The output voltage $v_2(t)$, which is the response to the rectangular pulse, is found in the same way as above:

$$\bar{V}_2(s) = \frac{s}{s + \frac{1}{\tau}} \cdot \frac{V(1 - e^{-sT_0})}{s};$$

$$\bar{V}_2(s) = \frac{V}{s + \frac{1}{\tau}} - \frac{Ve^{-sT_0}}{s + \frac{1}{\tau}};$$

thus,

$$v_2(t) = Ve^{-t/\tau}u(t) - Ve^{-(t-T_0)/\tau}u(t - T_0).$$

Hence, for $0 \leq t \leq T_0$,

$$v_2(t) = Ve^{-t/\tau},$$

and for $t > T_0$,

$$v_2(t) = V(e^{-T_0/\tau} - 1)e^{-(t-T_0)/\tau},$$

which is the result found from the differential-equation approach.

THE DC CIRCUIT SCHEME

To determine the transfer function of a circuit, we can use a scheme which greatly facilitates the solution of circuit problems. If we assume that the initial conditions are zero, that is, zero initial charge on all capacitors and zero initial current flowing through all inductors,* then we may think of the circuit as a direct current circuit with "resistive" elements:

Resistors	R
Capacitors	$1/sC$
Inductors	sL

These may also be referred to as generalized impedances. The circuit elements are represented by these generalized impedances and the circuit is solved as a dc voltage divider. The transfer function is just the relation between the input and output voltages, as before.

---

* Nonzero initial conditions may be treated as effective current and voltage sources. The superposition theorem may them be used to obtain the result for a real source plus nonzero initial conditions.

THE DC CIRCUIT SCHEME

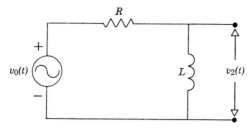

FIGURE 2-5. RL differentiating circuit.

As an example, consider the RL circuit shown in Fig. 2-5. Using the dc circuit scheme, we may immediately obtain the transfer function:

$$\bar{T}(s) = \frac{R_L}{R_R + R_L} = \frac{sL}{R + sL},$$

or

$$\bar{T}(s) = \frac{s\tau}{1 + s\tau} = \frac{s}{s + (1/\tau)},$$

with a time constant τ defined by $\tau = L/R$. This circuit is a differentiating circuit, since the transfer function written in terms of the time constant is identical to that of the RC differentiator studied before. One immediate advantage of this approach is that it allows us to see the similarity of circuits composed of quite different components, once we have obtained the transfer functions. The dc circuit scheme is based on the algebraic nature of the equations that result when one has used the Laplace transform to transform the time-varying equations encountered in a differential-equations approach to the complex s plane.

Another helpful result is that by replacing s by $j\omega$ in the transfer function and rationalizing the result, one obtains the gain $G(\omega)$ of the circuit at a frequency $f = \omega/2\pi$. Let us illustrate this with the above RL differentiator:

$$\bar{T}(s) = \frac{s\tau}{1 + s\tau};$$

$$\bar{T}(j\omega) = \frac{j\omega\tau}{1 + j\omega\tau}.$$

Rationalize the result by multiplying top and bottom by $(1 - j\omega\tau)$:

$$\bar{T}(j\omega) = \frac{j\omega\tau(1 - j\omega\tau)}{(1 + j\omega\tau)(1 - j\omega\tau)}$$

$$= \frac{(\omega\tau)^2}{1 + (\omega\tau)^2} + \frac{j\omega\tau}{1 + (\omega\tau)^2}$$

PASSIVE ELEMENTS: LAPLACE TRANSFORMS

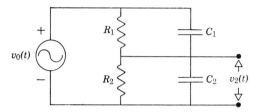

FIGURE 2-6. Compensated attenuator.

Express the result in terms of a magnitude and an angle, the magnitude being the gain $G(\omega) = |v_2(t)|/V$, and the angle being the phase of the output at frequency ω. Therefore,

$$G(\omega) = \frac{\omega\tau}{[1 + (\omega\tau^2)]^{1/2}}.$$

Define the phase angle relative to the complex axis, $\Delta\theta$, and then, once it has been calculated, transform to the angle $\Delta\phi$ expressed relative to the real axis by $\Delta\theta + \Delta\phi = \pi/2$. Therefore, the angle $\Delta\phi$, defined in the same way as in Chapter 1, becomes

$$\Delta\phi = \pi/2 - \tan^{-1}(\omega\tau)$$

These are the same results that we obtained in Chapter 1 for the *RC* differentiator. Thus, using the dc circuit scheme, we can obtain the same information that we found in the first chapter by a differential-equations approach but with far greater ease.

Once again, let us consider the compensated attenuator (Fig. 2-6):

$$\bar{T}(s) = \frac{\dfrac{1}{\dfrac{1}{R_2} + sC_2}}{\dfrac{1}{\dfrac{1}{R_1} + sC_1} + \dfrac{1}{\dfrac{1}{R_2} + sC_2}}$$

$$= \frac{R_2}{R_1 + R_2}\left(\frac{1 + s\tau_1}{1 + s\tau_\|}\right),$$

where

$$\tau_1 = R_1 C_1,$$
$$\tau_\| = R_\| C_\|,$$
$$\frac{1}{R_\|} = \frac{1}{R_1} + \frac{1}{R_2},$$
$$C_\| = C_1 + C_2.$$

THE DC CIRCUIT SCHEME

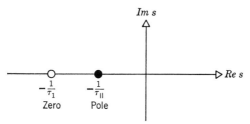

FIGURE 2-7. Pole-zero plot of the transfer function of a compensated attenuator.

The transfer function thus has a pole at $s = -1/\tau_\|$ and a zero at $s = -1/\tau_1$. Let us determine the response of the circuit to a step function of voltage:

$$\bar{V}_2(s) = \bar{T}(s)\bar{V}_0(s)$$

$$= \left(\frac{R_2}{R_1 + R_2}\right)\frac{V}{\tau_\|}\left[\frac{1}{s\left(s + \dfrac{1}{\tau_\|}\right)} + \frac{\tau_1}{s + \dfrac{1}{\tau_\|}}\right].$$

We can now look up the inverse transforms in a table or attempt to reduce them all to the exponential transform $1/(s + a)$. The last term is in this form, but the first is not. We can use *partial fractions* to put it in the proper form. Set

$$\frac{1}{s\left(s + \dfrac{1}{\tau_\|}\right)} = \frac{A}{s} + \frac{B}{s + \dfrac{1}{\tau_\|}};$$

$$1 = A\left(s + \frac{1}{\tau_\|}\right) + Bs.$$

If $s = 0$,

$$1 = \frac{A}{\tau_\|} \quad \text{and} \quad A = \tau_\|.$$

If

$$s = -\frac{1}{\tau_\|},$$

$$1 = -\frac{B}{\tau_\|} \quad \text{and} \quad B = -\tau_\|.$$

PASSIVE ELEMENTS: LAPLACE TRANSFORMS

Thus,

$$\frac{1}{s\left(s + \frac{1}{\tau_\|}\right)} = \tau_\| \left[\frac{1}{s} - \frac{1}{s + \frac{1}{\tau_\|}}\right];$$

$$\bar{V}_2(s) = \frac{R_2}{R_1 + R_2} V \left[\frac{1}{s} - \frac{\left(1 - \frac{\tau_1}{\tau_\|}\right)}{s + \frac{1}{\tau_\|}}\right];$$

$$v_2(t) = \frac{R_2}{R_1 + R_2} V \left[1 - \left(1 - \frac{\tau_1}{\tau_\|}\right) e^{-t/t_\|}\right].$$

The above solution involved mostly algebra and certainly was easier to carry out than the differential-equation approach.

Examination of the poles and zeros of the transfer function in the complex plane provides an interesting pastime (Fig. 2-7). The positions in the complex plane of the pole and zero may be adjusted by changing the parameters of the circuit, that is, the resistances and capacitances. In particular, the pole and zero may be made to overlap so that they "annihilate" each other, giving an s-independent transfer function. This technique is called *pole-zero cancellation*. Thus,

$$\bar{T}(s) = \frac{R_2}{R_1 + R_2} \left(\frac{1 + s\tau_1}{1 + s\tau_\|}\right) = \frac{R_2}{R_1 + R_2}$$

for $\tau_1 = \tau_\|$, or $R_1 C_1 = R_2 C_2$, the compensation condition. The output voltage is just reduced by the factor $R_2/(R_1 + R_2)$. We might remark at this point that the differentiating circuit may be made into a blocking circuit by moving the pole $s = -1/\tau$ toward the origin, where it will tend to be annihilated by the zero. This occurs for large time constant τ and gives $\bar{T}(s) = 1$.

So far, the poles we have considered have all been on the negative real axis and have lead to decaying exponentials. The series RCL circuit gives us the chance to examine the situation when the poles occur off the real axis. This time we take the output across the resistance (Fig. 2-8). The transfer function is

$$\bar{T}(s) = \frac{R}{\frac{1}{sC} + sL + R}$$

$$= \frac{R}{L} \left[\frac{s}{s^2 + \frac{R}{L} s + \frac{1}{LC}}\right].$$

54

THE DC CIRCUIT SCHEME

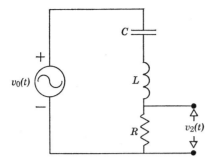

FIGURE 2-8. Series *RCL* circuit.

This transfer function has a zero at the origin and poles at the points in the complex plan where the denominator goes to zero. We factor the denominator:

$$s^2 + \frac{R}{L}s + \frac{1}{LC} = (s - \alpha_1)(s - \alpha_2) = 0;$$

$$\alpha_{1,2} = -\frac{R}{2L} \pm \frac{1}{2}\left[\frac{R^2}{L^2} - \frac{4}{LC}\right]^{1/2} = -\gamma \pm j\omega_1.$$

The response to a voltage step input is

$$\bar{V}_2(s) = \frac{RV}{L}\left[\frac{1}{(s - \alpha_1)(s - \alpha_2)}\right],$$

or since

$$(s - \alpha_1)(s - \alpha_2) = (s + \gamma - j\omega_1)(s + \gamma + j\omega_1)$$
$$= (s + \gamma)^2 + \omega_1^2,$$

$$\bar{V}_2(s) = \frac{RV}{L}\left[\frac{1}{(s + \gamma)^2 + \omega_1^2}\right].$$

Looking up this transform in the table gives us

$$v_2(t) = \frac{RV}{\omega_1 L} e^{-\gamma t} \sin \omega_1 t.$$

Looking at the pole-zero plot (Fig. 2-9), we see that the poles are complex conjugates of each other. This must be the case if we are to have an output voltage that is real.

We also note that the distance in the complex plane ω_1 from the real axis gives the frequency of *ringing;* and the distance γ along the negative real axis from the origin is a measure of the *damping* of the oscillation (Fig. 2-10).

PASSIVE ELEMENTS: LAPLACE TRANSFORMS

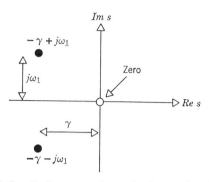

FIGURE 2-9. Pole-zero plot of of a series *RCL* circuit.

One may get rid of the oscillation in the output by moving the poles to the real axis, that is, by setting $\omega_1 = 0$. Then one has a double pole, and

$$\bar{V}_2(s) = \frac{RV}{L}\left[\frac{1}{(s+\gamma)^2}\right].$$

From the tables this gives

$$v_2(t) = \frac{RV}{L}te^{-\gamma t}.$$

This is the *critically damped* case discussed in Chapter 1. As γ is made larger than ω_0, the double pole splits, with one pole moving toward the origin and the other outward along the negative real axis. The resulting solution for

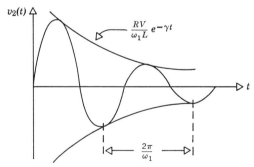

FIGURE 2-10. Output voltage of a series *RCL* circuit for an input step function. The output is taken across the resistor and the circuit is underdamped.

TRANSFORMERS

$v_2(t)$ is the sum of two decaying exponentials and yields the result we had before:

$$v_2(t) = \frac{RV}{\beta L} e^{-\gamma t} \sinh \beta t.$$

The path in the complex plane over which a pole moves as we vary some circuit parameter is referred to as a *pole trajectory*.

TRANSFORMERS

The use of transformers in circuits to change amplitude and impedance levels, to invert signals, or to perform ac coupling between elements while insuring dc isolation has the advantages of simplicity and stability. However, transformers will, in general, distort the transferred signal, and an analysis of the shaping effects can be gained by finding the voltage transfer function of the device. The approach to a device such as a transformer must be to find the simplest model, using realistic approximations, that gives the required information, since exact treatments are almost never feasible. We will consider an idealized transformer, consisting of a primary winding of N_1 turns having an inductance L_1, a secondary of N_2 turns with an inductance L_2, and a mutual inductance M (Fig. 2-11). The dot above (below) the coils indicates that the sense of the windings results in an output voltage of the same (opposite) sign as the input voltage. If a core is used, we will assume zero hysteresis and eddy current losses and linear permeability. Then

$$v_1(t) = L_1 \frac{di_1}{dt} + M \frac{di_2}{dt};$$

$$v_2(t) = M \frac{di_1}{dt} + L_2 \frac{di_2}{dt};$$

$$v_2(t) = -i_2(t)R_2.$$

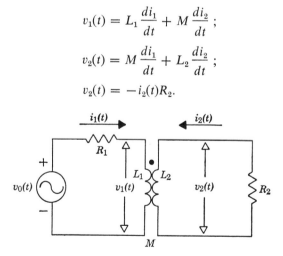

FIGURE 2-11. Schematic of a transformer.

PASSIVE ELEMENTS: LAPLACE TRANSFORMS

The proportionality constant M relating the voltage across the coil in one circuit due to a change in current in the other is called the *mutual inductance*. The constant can be either positive or negative depending on the sense of the windings. We can define a coupling constant k through the relation

$$k^2 = \frac{M^2}{L_1 L_2}.$$

If the permeability is high and the magnetic fields are essentially restricted to the core, then

$$\frac{L_2}{L_1} \simeq \frac{N_2^2 G_2}{N_1^2 G_1} \simeq \left(\frac{N_2}{N_1}\right)^2 = n^2$$

for identical geometrical factors, G_1 and G_2, of the two windings. N_1 and N_2 are the number of turns on each coil. To transform the above equations, we have to first find the Laplace transform of a time derivative of a function:

$$\mathscr{L}\left[\frac{df}{dt}\right] = \int_0^\infty e^{-st} \left(\frac{df}{dt}\right) dt.$$

Using integration by parts:

$$\int_0^\infty e^{-st} \frac{df}{dt} dt = e^{-st} f \Big|_0^\infty + s \int_0^\infty e^{-st} f \, dt,$$

or

$$\mathscr{L}\left[\frac{df}{dt}\right] = s\bar{F}(s) - f(0).$$

In most of our applications, we will assume that the function is initially at zero, and neglect $f(0)$. The transformer equations then become

$$\bar{V}_1(s) = L_1 s \bar{I}_1(s) + M s \bar{I}_2(s);$$
$$\bar{V}_2(s) = M s \bar{I}_1(s) + L_2 s \bar{I}_2(s);$$
$$\bar{V}_2(s) = -R_2 \bar{I}_2(s).$$

These equations can be easily solved algebraically for the required functional dependences:

$$\bar{V}_2(s) = -R_2 \left\{ \frac{M\bar{V}_1(s) - L_1 \bar{V}_2(s)}{s(M^2 - L_1 L_2)} \right\}.$$

Using the definition $k^2 = M^2/L_1 L_2$, this becomes

$$\bar{V}_2(s)\left\{1 + \frac{R_2}{sL_2} \frac{1}{(1-k^2)}\right\} = \bar{V}_1(s) \frac{R_2 M}{sL_1 L_2} \frac{1}{(1-k^2)}.$$

TRANSFORMERS

The voltage transfer function is found to be:

$$T = \frac{\bar{V}_2(s)}{\bar{V}_1(s)} = \frac{M}{L_1}\left\{\frac{1}{1 + s\dfrac{L_2}{R_2}(1 - k^2)}\right\}.$$

A word of caution is necessary at this point. Anticipating a more detailed treatment of the transfer function occurring later in this chapter, it must be noted that the transfer function is in essence an approximation based partly on a high impedance load. This condition is well satisfied during the leading edge of a step function, since the high frequencies involved result in a large impedance in the transformer's secondary; but away from the leading edge, the approximation breaks down. Therefore, this analysis will treat only the fast time response of a transformer.

Writing $\tau_2 = L_2/R_2$, and defining $\tau = \tau_2(1 - k^2)$, the transfer function is just that of the familiar integrating circuit, with a gain M/L_1:

$$T = \frac{M}{L_1}\left(\frac{1}{1 + s\tau}\right) = \frac{M}{L_1}\left(\frac{1/\tau}{s + 1/\tau}\right).$$

Note that the gain does not depend on the inductance of the secondary nor the load R_2. The pole-zero plot is that of a decaying exponential (Fig. 2-12).

Note that by varying the coefficient of coupling k, one can move the pole out to ∞ along the negative real axis. This means that the rise time of the device is very small and the device will follow the input perfectly. This limiting case is that of an *ideal transformer*, given by $k = 1$. The resulting voltage gain is

$$T = \frac{M}{L_1} = \frac{\sqrt{L_1 L_2}}{L_1} = \sqrt{\frac{L_1 L_2}{L_1^2}} \quad \text{for} \quad \frac{M^2}{L_1 L_2} = 1.$$

Then

$$T = \sqrt{\frac{L_2}{L_1}} \simeq \sqrt{n^2} = n, \text{ the turns ratio.}$$

To find the shaping effects in a non-ideal transformer (Fig. 2-13), consider a step function input:

$$\bar{V}_2(s) = \frac{M}{L_1}\left(\frac{1}{1 + s\tau}\right)\frac{V_0}{s},$$

or

$$\bar{V}_2(s) = \frac{V_0 M}{L_1}\left(\frac{1}{s} - \frac{1}{s + \dfrac{1}{\tau}}\right).$$

PASSIVE ELEMENTS: LAPLACE TRANSFORMS

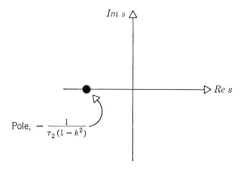

FIGURE 2-12. Pole-zero plot of the transfer function of a transformer.

Transforming back to the time domain,

$$v_2(t) = \frac{V_0 M}{L_1}(1 - e^{-t/\tau}).$$

Since one of the uses of a transformer is to match the impedance of a source to that of a load, we should at this point calculate the input impedance of a transformer. Given the definition of the input impedance

$$Z_{in}(s) = \frac{\bar{V}_1(s)}{\bar{I}_1(s)},$$

then

$$-R_2 \bar{I}_2(s) = s[(M\bar{I}_1(s) + L_2 \bar{I}_2(s)];$$
$$\bar{I}_2(s)(sL_2 + R_2) = -sM\bar{I}_1(s);$$
$$\bar{V}_1(s) = sL_1 \bar{I}_1(s) + sM\left(\frac{-sM\bar{I}_1(s)}{R_2 + sL_2}\right);$$
$$Z_{in}(s) = \frac{\bar{V}_1(s)}{\bar{I}_1(s)} = sL_1 - \frac{s^2 M^2}{R_2 + sL_2}.$$

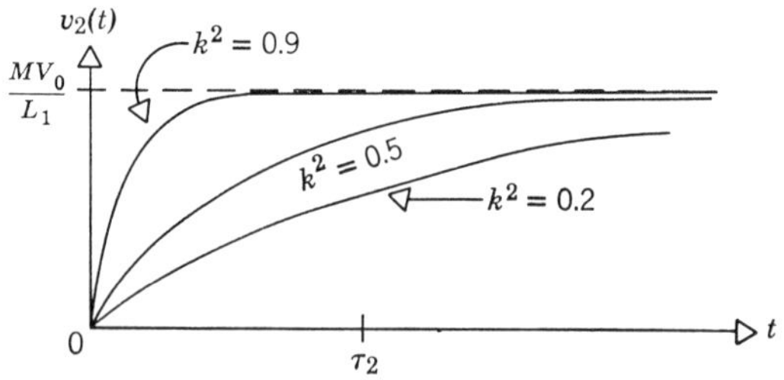

FIGURE 2-13. Output voltage of a transformer as a function of the coupling constant for an input step function.

TRANSFORMERS

Thus,

$$Z_{in}(s) = sL_1 \left[\frac{1 + s\tau_2(1 + k^2)}{1 + s\tau_2} \right];$$

$$Z_{in} \simeq \frac{L_1 R_2}{L_2} \simeq \frac{R_2}{n^2}.$$

Using the results of Problem 2-9, the output impedance under the same conditions becomes

$$Z_{out} \simeq \frac{L_2}{L_1} R_1 \simeq n^2 R_1.$$

We see, then, that we may match impedances by adjusting the turns ratio n so that $Z_{in} = R_1$, or $n = \sqrt{R_2/R_1}$.

To make our model more realistic, we must consider the effective capacitance of the transformer, since an electric field exists turn to turn on the windings. As an example of the type of effects which occur, let us consider first the effect of the capacitance on the output side (Fig. 2-14). As before,

$$V_1(s) = L_1 s \bar{I}_1(s) + Ms \bar{I}_2(s);$$
$$\bar{V}_2(s) = Ms \bar{I}_1(s) + L_2 s \bar{I}_2(s).$$

But now,

$$\bar{V}_2(s) = -I_2 \left(\frac{1}{\frac{1}{R_2} + sC_2} \right) = \frac{-I_2 R_2}{1 + sR_2 C_2}.$$

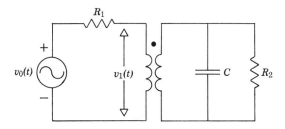

FIGURE 2-14. Schematic of a transformer including capacitance.

PASSIVE ELEMENTS: LAPLACE TRANSFORMS

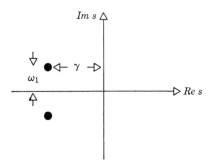

FIGURE 2-15. Pole-zero plot of the transfer function of a transformer including capacitance.

Since we are working only with algebra, we can replace R_2 by $R_2/(1 + sR_2C_2)$ in our previous equations:

$$\bar{T}(s) = \frac{M}{L_1}\left(\frac{1}{1 + s\left[\dfrac{L_2(1 + sR_2C_2)(1 - k^2)}{R_2}\right]}\right);$$

$$\bar{T}(s) = \frac{MR_2}{L_1}\left(\frac{1}{R_2 + L_2(1 - k^2)s + L_2(1 - k^2)R_2C_2s^2}\right);$$

$$\bar{T}(s) = \frac{M}{L_1}\left(\frac{1}{L_2C_2(1 - k^2)}\right)\left(\frac{1}{s^2 + \dfrac{s}{R_2C_2} + \dfrac{1}{L_2C_2(1 - k^2)}}\right).$$

The poles are at $\alpha_{1,2} = -\gamma \pm j\omega_1$ (Fig. 2-15):

$$\alpha_{1,2} = -\frac{1}{2R_2C_2} \pm \frac{1}{2}\left[\frac{1}{(R_2C_2)^2} - \frac{4}{L_2C_2(1 - k^2)}\right]^{1/2}.$$

For $1/L_2C_2(1 - k^2) > 1/(2R_2C_2)^2$, ω_1 is real and $j\omega_1$ is thus imaginary, resulting in an oscillation damped by the time constant γ.

Consider an input step function. Then

$$\bar{V}_2(s) = \bar{T}(s)\bar{V}_0(s) = \frac{M}{L_1}\frac{\omega_0^2}{(1 - k^2)}\left(\frac{1}{(s - \alpha_1)(s - \alpha_2)}\right)\frac{V_0}{s}$$

$$\alpha_1 = -\gamma + j\omega_1$$

$$\alpha_2 = -\gamma - j\omega_1$$

$$\omega_0^2 = \frac{1}{L_2C_2}$$

$$\bar{V}_2(s) = \frac{MV_0}{L_1}\frac{\omega_0^2}{(1 - k^2)}\left(\frac{1}{s(s - \alpha_1)(s - \alpha_2)}\right).$$

RF CIRCUITRY

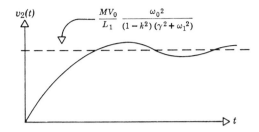

FIGURE 2-16. Output voltage of a transformer including capacitance.

Looking up this function and finding its inverse transform (Fig. 2-16), we have

$$v_2(t) = \frac{MV_0}{L_1} \frac{\omega_0^2}{(1-k^2)(\gamma^2 + \omega_1^2)} \left\{ 1 - e^{-\gamma t}\left(\cos \omega_1 t + \frac{\gamma}{\omega_1} \sin \omega_1 t\right)\right\}.$$

Then, as $k^2 \Rightarrow 1$, $\gamma^2 + \omega_1^2 \Rightarrow \omega_1^2 \Rightarrow \omega_0^2/(1-k^2)$;

$$v_2(t) \Rightarrow \frac{MV_0}{L_1} = nV_0.$$

RF CIRCUITRY

When considering the response of circuits to sinusoidal signals, we must consider limitations of the various models of the components. At radio frequencies and above, we may no longer consider a resistor as just having resistance, but must also consider the series inductance and parallel distributed capacitance. Our circuit analysis, however, is not changed, but will become much more complicated as we increase the complexity of each component. There is another way in which the simple lumped-parameter models break down, and this is the assumption that all electromagnetic fields are contained within the component itself. At high frequencies, the various components—resistors, capacitors, and inductors—radiate electromagnetic radiation; thus, they may be coupled with other components, causing feedback and inherent oscillations of any given circuit, for example, a high-frequency amplifier.

PASSIVE ELEMENTS: LAPLACE TRANSFORMS

One necessary addition to circuitry when working at higher frequencies is electromagnetic shielding. Various plates at ground potential may be placed around the components so that different parts of the circuits are isolated from each other. If sufficient care is taken, the high-frequency amplifier may be made quite stable and will not oscillate.

MORE ABOUT THE TRANSFER FUNCTION

This might be a good place to discuss the nature of the transfer function in a bit more depth. Basically, the voltage transfer function relates an input voltage to an output voltage, which is precisely the same thing done by Thevenin's theorem. We can draw the circuits implied by the two equations and compare them (Fig. 2-17).

$\bar{V}_S(s)$ and $\bar{Z}_S(s)$ are the Thevenin's theorem parameters of the source supplying the input voltage $\bar{V}_0(s)$ to the circuit under study, while $\bar{V}_L(s)$ and $\bar{Z}_L(s)$ are the Thevenin's theorem parameters of the load present at the output terminals of the circuit under study, across which appears $\bar{V}_2(s)$. The Thevenin's theorem parameters for the circuit under study are $\bar{Z}_{in}(s)$ and $\bar{V}_{in}(s)$ for the input leg, and the corresponding values for the output leg are similarly indicated. For simplicity, assume that $\bar{V}_{in}(s)$ and $\bar{V}_L(s)$ are zero, as is usually the case, since these represent back voltages in the circuit independent of the impedances present.

How do these two formulations compare? Assume that the voltage $\bar{V}_0(s)$ is present before the circuit under study is hooked up to it, in which case it equals $\bar{V}_S(s)$ since no current is flowing through $\bar{Z}_S(s)$. If the circuit under study draws significant current, then $\bar{V}_0(s)$ will no longer equal $\bar{V}_S(s)$. This would certainly modify our ideas concerning the nature of the transfer function, since voltage alone would not really be sufficient to describe the resultant behavior. Therefore, if we wish to use the transfer function as the *sole* circuit parameter, we must assume that $\bar{Z}_{in}(s)$ is much greater than $\bar{Z}_S(s)$ at all frequencies. That is why we use an ideal voltage source for all inputs, thus setting $\bar{Z}_S(s) = 0$.

Likewise, at the output terminals, it would not do us much good to have $\bar{V}_2(s) = \bar{T}(s)\bar{V}_0(s)$ if significant current flowing into $\bar{Z}_L(s)$ would reduce $\bar{V}_2(s)$ by $\bar{I}_2(s)\bar{Z}_{out}(s)$. Therefore, we must assume, if $\bar{T}(s)$ is to be our *sole* circuit parameter, that $\bar{Z}_{out}(s)$ is much less than $\bar{Z}_L(s)$ at all frequencies. That is why we write $v_2(t)$ as though it possessed infinite impedance ($i_2(t) = 0$).

Thus, the transfer function should, in general, be supplemented by the input and output impedances of the circuit under study. In this case, the Thevenin's theorem equivalent voltage is just $\bar{T}(s)\bar{V}_0(s)$, and the circuit reduces to that shown in Fig. 2-18.

MORE ABOUT THE TRANSFER FUNCTION

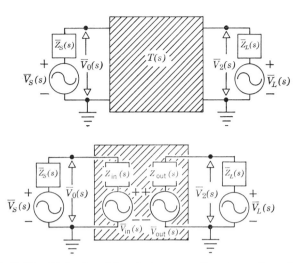

FIGURE 2-17. Relationship between the transfer function and Thevenin's theorem

In many cases, especially those involving high input impedances, the transfer function alone is sufficient to describe the behavior of a circuit. Also, in situations in which $\bar{Z}_{in}(s)$ and $\bar{Z}_{out}(s)$ are mainly resistive, frequency dependence will reside in the transfer function alone, and the output wave form can be quickly obtained without doing the more exact calculation demanded by the Thevenin's theorem.

By using Kirchhoff's laws and the dc circuit scheme, the reader may show that the voltage transfer function and the input and output impedances may be identified in the general case from the following expression:

$$\bar{V}_2(s) = \bar{T}(s) \frac{\bar{Z}_{in}(s)}{\bar{Z}_{in}(s) + \bar{Z}_S(s)} \bar{V}_S(s) - \bar{I}_2(s)\bar{Z}_{out}(s).$$

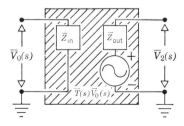

FIGURE 2-18. A somewhat simplified equivalent circuit based on the transfer function and Thevenin's theorem.

PASSIVE ELEMENTS: LAPLACE TRANSFORMS

APPENDIX

Short Table of Laplace Transforms

$f(t)$	$\bar{F}(s)$
$f(t)$	$\int_0^\infty f(t)e^{-st}\,dt$
$u(t)$	$1/s$
df/dt	$s\bar{F}(s) - f(0)$
$\int_0^t f(t')\,dt'$	$\bar{F}(s)/s$
$\int_0^t f_1(t-t')f_2(t')\,dt'$	$\bar{F}_1(s)\bar{F}_2(s)$
$f(t/a)$	$a\bar{F}(as)$
$f(t-a)u(t-a)$	$e^{-sa}\bar{F}(s)$
$1/\sqrt{\pi t}$	$1/\sqrt{s}$
e^{-at}	$1/(s+a)$
$1/(b-a)[e^{-at} - e^{-bt}]$	$1/(s+a)(s+b)$
$1/(a-b)[ae^{-at} - be^{-bt}]$	$s/(s+a)(s+b)$
$\dfrac{1}{(b-a)(c-a)}e^{-at} + \dfrac{1}{(a-b)(c-b)}e^{-bt} + \dfrac{1}{(a-c)(b-c)}e^{-ct}$	$1/(s+a)(s+b)(s+c)$
te^{-at}	$1/(s+a)^2$
$\dfrac{1}{(n-1)!}t^{n-1}e^{-at}$	$1/(s+a)^n$
$[1 - (1+at)e^{-at}]/a^2$	$1/s(s+a)^2$
$e^{-at}(1-at)$	$s/(s+a)^2$
$\sin(at)$	$a/(s^2+a^2)$
$\cos(at)$	$s/(s^2+a^2)$
$\sinh(at)$	$a/(s^2-a^2)$
$\cosh(at)$	$s/(s^2-a^2)$
$\dfrac{1}{b}e^{-at}\sin(bt)$	$\dfrac{1}{(s+a)^2+b^2}$
$e^{-at}f(t)$	$\bar{F}(s+a)$
$\delta(t)$	1
$t^n/(n!)$	$1/s^{n+1}$

PROBLEMS

1. Using integration by parts, show that the Laplace transform of the definite integral is
$$\mathscr{L}\left[\int_0^t f(t)\,dt\right] = \frac{\bar{F}(s)}{s}.$$
What is the Laplace transform of the indefinite integral $\int f(t)\,dt$?

2. Show by the method of the Laplace transform that if the capacitor in an *RC* differentiating circuit has an initial charge q_0, the output voltage is given by
$$v_2(t) = \left(V - \frac{q_0}{c}\right) e^{-t/\tau}.$$

3. Find the transfer function for the following circuits:

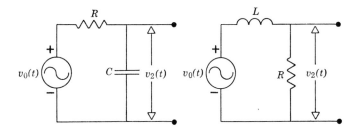

Comment on your results. If one had a black box whose transfer function was that found above, could one tell if the actual components were *RC* or *RL*? If you had other passive elements at your disposal, how could you determine the actual configuration, assuming that the circuit in the box had only two elements?

4. Find the input impedance of an *RC* differentiating circuit. Is it the same as that of an *RL* differentiating circuit? What assumptions do you have to make about the load?

5. Find the input and output impedances of a compensated attenuator. Assume resistive impedances in the voltage source and the load. At the compensation condition, are these impedances frequency independent? Comment.

6. Find the input and output impedances of *RC* and *RL* integrating circuits.

7. Using the dc circuit scheme, verify that the gain of an integrating circuit is that derived in Chapter 1.

8. At the -3-dB point of a low-pass circuit based on resistors and capacitors, how do the values of the input and output impedance relate to their values at low frequency?

9. Show that the output impedance of a perfect transformer is
$$Z_{\text{out}} \simeq n^2 R_1.$$

PASSIVE ELEMENTS: LAPLACE TRANSFORMS

10. Consider a parallel *RCL* circuit. Is the concept of a transfer function useful in this case? Discuss. For the case of such a circuit driven by a current source, a *transfer impedance* function is appropriate. How should such a function be defined and what would it be for this circuit?

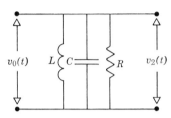

11. We can define a *current transfer function* in exactly the same manner as our voltage transfer function. In terms of an input current $\bar{I}_1(s)$, show that the output current, for the case of an ideal transformer, is

$$\bar{I}_2(s) = \bar{I}_1(s)/n \qquad n = N_1/N_2, \text{ the turns ratio.}$$

12. Show that for a series *RCL* tuned circuit the input impedance, in response to a sinusoidal signal of frequency ω near resonance $\omega_0 = 1/\sqrt{LC}$, is to first order in $\Delta\omega/\omega_0$, given by

$$\hat{Z}_{\text{in}} = R\left[1 + j2Q\frac{\Delta\omega}{\omega_0}\right],$$

where the Q for the series circuit is given by

$$Q = \frac{\omega_0 L}{R}.$$

13. What is the exact input impedance of a series *RCL* circuit? Graph its behavior versus frequency for the case in which the output voltage is taken across the resistor. Assume a resistive source and load, each with 100 ohms resistance, 0.1 h inductance, 10^{-6} farad capacitance, and 10 ohms circuit resistance. What is the output impedance at resonance, and how does it vary with frequency?

14. For the same circuit parameters, find the gain of a series *RCL* circuit versus frequency.

15. Find the transfer function for the circuit shown in Fig. 2-1.

16. Design a system to perform a phase shift of 45° at 10^4 Hz using resistive and capacitive elements. Would there be any advantage in using an *RL* circuit over an *RC* circuit or using a differentiator rather than an integrator? Discuss situations in which one is preferable over another.

EXPERIMENT 2

17. Consider the following circuit.

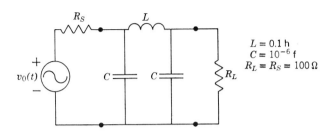

This is called a π section low-pass filter.
(a) Derive the transfer function for the circuit. Plot $G(\omega)$.
(b) What are the input and output impedances of the circuit?
(c) Discuss the utilization of such a circuit for filtering purposes in comparison to a simple RC filter.
(d) Look up in one of the bibliographic references the nature of the ripple present in this circuit and in an RC filtering circuit.

Experiment 2
RCL Circuits

In the second experiment, we will consider the behavior of circuits containing resistors, capacitors, and inductors, and the resonant phenomena that occur when these elements are present in various combinations. Perhaps it was noticed in the first experiment that resonant behavior appeared even when capacitors and inductors were not explicitly present together in a circuit. The explanation lies in the fact that whenever there is an electric field, there is capacitance present, and whenever there is a magnetic field, inductance. Therefore, *all* circuits are RCL circuits to some degree, although in many cases, the effect of this "stray" capacitance and inductance is negligible. It might be informative to calculate the amount of stray capacitance that appears when 1 cm of wire of diameter 0.01 cm is 1 cm away from a grounded chassis. What is the impedance for a signal at 10 MHz? Negligible?

Equipment:

1. Oscilloscope
2. Sine- and square-wave generator
3. Capacitor, 0.001 mf

PASSIVE ELEMENTS: LAPLACE TRANSFORMS

4. Inductor, 0.1 h
5. Resistor, 1 kΩ

I. Transient response

   A. Set up the circuit shown below.

   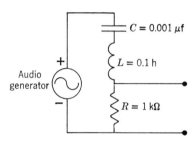

   1. Set the audio generator to 500~/sec, square wave.
   2. Connect 'scope to R. Adjust 'scope controls to give a fairly large pattern.

   B. Measure the period T of the *underdamped oscillation*.
   C. Measure $v_2(t)$ and $v_2(t+T)$ to obtain the logarithmic decrement, as defined in the text.
   D. Calculate γ and ω_1 from these measured values.
   E. Calculate ω_0, using the values of γ and ω_1.
   F. Compute the Q of the circuit from

   $$Q = \frac{\omega_0}{2\gamma}.$$

   G. Interchange R and C in the above diagram, and draw the observed waveform across C.

II. Steady-state response

   A. Using the same circuit as above, with the audio generator switched to the sine function, measure V_R, V_C, and V_L as functions of frequency. Take a band of frequencies such that a good resonance curve is obtained.
   B. Compute the values of conductance and susceptance at each of the above points.
   C. Plot the conductance and susceptance as functions of frequency.

EXPERIMENT 2

D. From the conductance curve, determine $\Delta\omega = |\omega_1 - \omega_2|$, where ω_1 = upper half-power point and ω_2 = lower half-power point. Then determine the Q of the circuit by

$$\frac{1}{Q} = \frac{\Delta\omega}{\omega_0}.$$

The value of ω_0 should be determined experimentally as well as calculated.

E. At resonance, measure the phases of V_R, V_C, and V_L with respect to the applied voltage, as follows:
   1. Run a jumper from the "hot" side of the signal generator to the Ext. Horiz. Input jack.
   2. Put the probe across V_R (V_C, V_L).
   3. With the Horiz. Display on Norm., set the amplitude of the signal to 6 cm.
   4. Switch the input selector to the open input, and the Horiz. Display to Ext.
   5. Using the Stability or Horiz. Input Atten. control, set the width of the line to 6 cm., and center the trace.
   6. Switch the input selector to the circuit under test to observe a Lissajous pattern (see Appendix to this experiment).
   7. Determine the x and y intercepts of the Lissajous pattern.

F. Determine ω such that $R = X$, where

$$X = X_L - X_C.$$

This determines the point where the phase shift is 45°.

   1. Set the sig. gen. to this frequency, and again measure the x and y intercepts of the Lissajous patterns obtained across R, L, and C, as above.
   2. By means of the formulas given in class, verify that the phases are as follows:
       a. At resonance,
          V_R has 0° phase shift
          V_L has +90° phase shift
          V_C has −90° phase shift
       b. At the ω determined above,
          V_R has +45° phase shift
          V_L has +45° phase shift
          V_C has −45° phase shift

PASSIVE ELEMENTS: LAPLACE TRANSFORMS

APPENDIX TO EXPERIMENT 2: PHASE MEASUREMENTS BY LISSAJOUS FIGURES

Assume the following inputs to the vertical and horizontal plates of an oscilloscope:

$$v_V = V_V \sin \omega t;$$
$$v_H = V_H \sin (\omega t + \phi).$$

The amplitudes on the face of the oscilloscope may be made equal by the use of attenuators or amplifiers. Then

$$y = A \sin \theta;$$
$$x = A \sin (\theta + \phi)$$

is a parametric representation of the resulting curve traced out by the electron beam. The parameter $\theta = \omega t$ may be eliminated to give the equation of the curve.

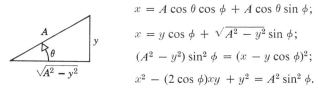

$$x = A \cos \theta \cos \phi + A \cos \theta \sin \phi;$$
$$x = y \cos \phi + \sqrt{A^2 - y^2} \sin \phi;$$
$$(A^2 - y^2) \sin^2 \phi = (x - y \cos \phi)^2;$$
$$x^2 - (2 \cos \phi)xy + y^2 = A^2 \sin^2 \phi.$$

This quadratic form may be diagonalized to find the principle axes of the ellipse. One way to do this is to perform a rotation of axes:

$$x = x' \cos \delta - y' \sin \delta;$$
$$y = x' \sin \delta + y' \cos \delta.$$

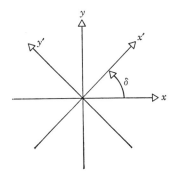

In terms of the new coordinates, the quadratic form becomes

$$(1 - \cos \phi \sin 2\delta)x'^2 + (1 + \cos \phi \sin 2\delta)y'^2 - (2 \cos \phi \cos 2\delta)x'y' = A^2 \sin^2 \phi.$$

The principle axes are defined to be those such that no $x'y'$ cross term appears in the quadratic form. This results in the condition

$$\cos 2\delta = 0,$$

or

$$\delta = \pm \frac{\pi}{4}.$$

APPENDIX TO EXPERIMENT 2

The quadratic form becomes

$$\frac{x'^2}{a^2} + \frac{y'^2}{b^2} = 1,$$

where

$$a = \frac{A \sin \phi}{\sqrt{1 \mp \cos \phi}};$$

$$b = \frac{A \sin \phi}{\sqrt{1 \pm \cos \phi}}.$$

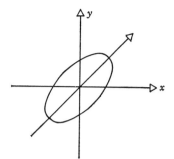

The phase difference ϕ may be found most easily by looking at the quadratic form in x, y space. It is to be noted that if $x = 0$,

$$y_0 = \pm A \sin \phi;$$

and if $y = 0$,

$$x_0 = \pm A \sin \phi.$$

Thus, A is measured and the point at which the ellipse crosses the y axis is measured and the phase shift ϕ computed from

$$\sin \phi = \frac{y_0}{A}.$$

TRANSMISSION AND DELAY LINES

3

THE LOSS-LESS TRANSMISSION LINE

It is frequently necessary to have various pieces of electronic equipment placed at some distance from each other. This means that the signals of interest must be transmitted over long cables or transmission lines. It is important, then, to have some model of a transmission line in order to determine what effect the line will have upon the signals. We first consider the simplest model, that is, the loss-less transmission line. We assume the line to consist of two conductors and to have a fixed capacitance per unit length C and a fixed inductance per unit length L. We represent an infinitesimal section of the line symbolically as shown in Fig. 3-1.

TRANSMISSION AND DELAY LINES

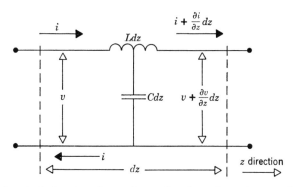

FIGURE 3-1. Infinitesimal section of a loss-less transmission line.

The change in voltage dv across the section of the line will be the back e.m.f. due to the inductance of the section:

$$dv = \frac{\partial v}{\partial z} dz = -(L\,dz)\frac{\partial i}{\partial t}.$$

The current in the upper conductor leaving this section of the line will be diminished by the charging current di diverted to the capacitance $C\,dz$. Hence,

$$di = \frac{\partial i}{\partial z} dz = -(C\,dz)\frac{\partial v}{\partial t}.$$

The equations which govern the spatial and time dependence of the voltage and current for the loss-less transmission line are, then, two coupled first-order partial differential equations with constant coefficients:

$$\frac{\partial v}{\partial z} = -L\frac{\partial i}{\partial t};$$

$$\frac{\partial i}{\partial z} = -C\frac{\partial v}{\partial t}.$$

We notice that differentiating the first equation with respect to z and the second with respect to t and combining yields the wave equation for the voltage $v(z, t)$:

$$\frac{\partial^2 v}{\partial z^2} = \frac{1}{\mu^2}\frac{\partial^2 v}{\partial t^2}.$$

The velocity of the waves is given by

$$\mu = \frac{1}{\sqrt{LC}}.$$

THE LOSS-LESS TRANSMISSION LINE

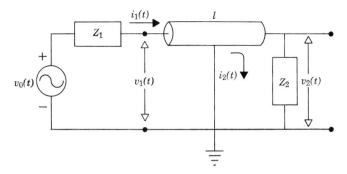

FIGURE 3-2. Schematic of a transmission line.

The two equations above may also be combined to give a wave equation for the current with the same velocity. Thus, there will exist traveling waves (both forward and backward) of voltage and current in the transmission line:

$$v(z, t) = v_{\leftarrow}\left(t + \frac{z}{\mu}\right) + v_{\rightarrow}\left(t - \frac{z}{\mu}\right);$$

$$i(z, t) = i_{\leftarrow}\left(t + \frac{z}{\mu}\right) + i_{\rightarrow}\left(t - \frac{z}{\mu}\right).$$

In practice, a transmission line will be connected to a voltage source at one end and a load at the other. Since the impedance of the line will usually turn out to be close to the impedance of the source and the load, we will have to use a full Thevenin's theorem formulation for the source and the load. Consider a line of length l shown schematically in Fig. 3-2.

To be able to work with general impedances, reactive as well as resistive, we use the Laplace transform method and transform the coupled partial differential equations with respect to time. Since

$$\bar{I}(s, z) = \int_0^\infty i(z, t)e^{-st}\, dt;$$

$$\bar{V}(s, z) = \int_0^\infty v(z, t)e^{-st}\, dt,$$

we have

$$\frac{\partial \bar{V}}{\partial z} = -sL\bar{I} + Li(z, 0);$$

$$\frac{\partial \bar{I}}{\partial z} = -sC\bar{V} + Cv(z, 0),$$

TRANSMISSION AND DELAY LINES

where we have used the Laplace transform of the time derivative of a function:

$$\mathscr{L}\left[\frac{\partial f}{\partial t}\right] = s\bar{F}(s) - f(0).$$

Combining the above partial differential equations gives us the second-order equation:

$$\frac{\partial^2 \bar{V}}{\partial z^2} - s^2 LC\bar{V} = -sLCv(z, 0) + L\frac{\partial i}{\partial z}(z, 0).$$

In general, we will be interested only in deviations from quiescent values of voltage and current, so the right-hand side of this equation may be set equal to zero.

We then have the exponential differential equation

$$\frac{\partial^2 \bar{V}}{\partial z^2} - s^2 LC\bar{V} = 0,$$

with solution

$$\bar{V}(s, z) = \bar{V}_{\leftarrow}(s)e^{sTz/l} + \bar{V}_{\rightarrow}(s)e^{-sTz/l},$$

where

$$T = l\sqrt{LC}$$

is the *transit time* down the line.

The time-dependent voltage is found from the inverse transform, which involves the use of the shifting theorem:

$$v(z, t) = v_{\leftarrow}\left(t + \frac{z}{\mu}\right) + v_{\rightarrow}\left(t - \frac{z}{\mu}\right).$$

From the symmetry of the coupled equations we see that a similar solution exists for the current:

$$\bar{I}(s, z) = \bar{I}_{\leftarrow}(s)e^{sTz/l} + \bar{I}_{\rightarrow}(s)e^{-sTz/l}.$$

In addition to the transit time of the line (or delay per unit length), there is one other parameter which is characteristic of the loss-less line. The ratio of the forward voltage wave to the forward current wave is a constant. Let us work with the Laplace transforms of the forward waves. We have the *forward voltage wave*

$$\bar{V}_{\rightarrow}(s, z) = \bar{V}_{\rightarrow}(s)e^{-sTz/l}$$

and the *forward current wave*

$$\bar{I}_{\rightarrow}(s, z) = \bar{I}_{\rightarrow}(s)e^{-sTz/l}.$$

TERMINATION

To prove the above statement, let us consider one of the first-order coupled partial differential equations. Substituting the general solutions into the first of these gives

$$\bar{V}_{\leftarrow}(s)\frac{sTe^{sTz/l}}{l} - \bar{V}_{\rightarrow}(s)\frac{sTe^{-sTz/l}}{l} = -sL[\bar{I}_{\leftarrow}(s)e^{sTz/l} + \bar{I}_{\rightarrow}(s)e^{-sTz/l}],$$

or

$$\left[\frac{T}{l}\bar{V}_{\leftarrow}(s) + L\bar{I}_{\leftarrow}(s)\right]e^{sTz/l} = \left[\frac{T}{l}\bar{V}_{\rightarrow}(s) - L\bar{I}_{\rightarrow}(s)\right]e^{-sTz/l}.$$

The only way in which this equation may be satisfied for all z is for the coefficients of the exponentials to be identically zero. Hence,

$$\frac{\bar{V}_{\rightarrow}(s)}{\bar{I}_{\rightarrow}(s)} = \frac{lL}{T} = \frac{lL}{l\sqrt{LC}} = \sqrt{\frac{L}{C}},$$

or

$$\frac{\bar{V}_{\rightarrow}(s,z)}{\bar{I}_{\rightarrow}(s,z)} = Z_0,$$

where

$$Z_0 = \sqrt{\frac{L}{C}}.$$

Z_0 is called the *characteristic impedance* of the line. We also have

$$\frac{\bar{V}_{\leftarrow}(s,z)}{\bar{I}_{\leftarrow}(s,z)} = -Z_0.$$

An important point to note here is that, although the characteristic impedance is s-independent and has dimensions of resistance, it is not a resistance in the true sense of the word. There is no dissipation of energy, no joule heating involved.

TERMINATION

Let us now evaluate the voltage and current at the end of the line ($z = l$). Since we are interested in deviations from quiescent values, we may use the dc circuit scheme:

$$\bar{V}_2(s, l) = \bar{I}_2(s, l)\bar{Z}_2(s);$$

$$\bar{V}_{\leftarrow}(s)e^{sT} + \bar{V}_{\rightarrow}(s)e^{-sT} = [\bar{I}_{\leftarrow}(s)e^{sT} + \bar{I}_{\rightarrow}(s)e^{-sT}]\bar{Z}_2(s).$$

TRANSMISSION AND DELAY LINES

Using the relations between the voltage and currrent waves, we may solve this equation for the ratio of the backward wave to the forward wave:

$$\frac{\bar{V}_{\leftarrow}(s)e^{sT}}{\bar{V}_{\rightarrow}(s)e^{-sT}} = \bar{\eta}_2(s),$$

where

$$\bar{\eta}_2(s) = \frac{\bar{Z}_2(s) - Z_0}{\bar{Z}_2(s) + Z_0}$$

is the *reflection coefficient at the end of the line*. At the front of the line ($z = 0$), we have

$$\bar{V}_1(s,0) = \bar{V}_0(s) - \bar{I}_1(s, 0)\bar{Z}_1(s);$$

or

$$\bar{V}_{\leftarrow}(s) + \bar{V}_{\rightarrow}(s) = \bar{V}_0(s) - [\bar{I}_{\leftarrow}(s) + \bar{I}_{\rightarrow}(s)]\bar{Z}_1(s),$$

$$\bar{V}_{\rightarrow}(s) = \bar{A}(s)\bar{V}_0(s) + \bar{\eta}_1(s)\bar{V}_{\leftarrow}(s),$$

where

$$\bar{\eta}_1(s) = \frac{\bar{Z}_1(s) - Z_0}{\bar{Z}_1(s) + Z_0}$$

is the *reflection coefficient at the front of the line* and

$$\bar{A}(s) = \frac{Z_0}{\bar{Z}_1(s) + Z_0}$$

is the *voltage attenuation ratio*. Solving the above equations for $\bar{V}_{\rightarrow}(s)$ and $\bar{V}_{\leftarrow}(s)$ results in

$$\bar{V}_{\rightarrow}(s) = \frac{\bar{A}(s)}{1 - \bar{\eta}_1(s)\bar{\eta}_2(s)e^{-2sT}} [\bar{V}_0(s)]$$

and

$$\bar{V}_{\leftarrow}(s) = \frac{\bar{A}(s)\bar{\eta}_2(s)e^{-2sT}}{1 - \bar{\eta}_1(s)\bar{\eta}_2(s)e^{-2sT}} [\bar{V}_0(s)].$$

We may now determine the "voltages" at the end and at the front of the line:

$$\bar{V}_2(s, l) = \bar{V}_{\leftarrow}(s)e^{sT} + \bar{V}_{\rightarrow}(s)e^{-sT}, \quad \text{or} \quad \bar{V}_2(s, l) = \bar{T}_2(s)\bar{V}_0(s),$$

where

$$\bar{T}_2(s) = \frac{\bar{A}(s)[1 + \bar{\eta}_2(s)]e^{-sT}}{1 - \bar{\eta}_1(s)\bar{\eta}_2(s)e^{-2sT}}$$

is the *voltage transfer function at the end of the line*. Also, we have

$$\bar{V}_1(s, 0) = \bar{V}_{\leftarrow}(s) + \bar{V}_{\rightarrow}(s),$$

or

$$\bar{V}_1(s, 0) = \bar{T}_1(s)\bar{V}_0(s),$$

TERMINATION

where

$$\bar{T}_1(s) = \frac{\bar{A}(s)[1 + \bar{\eta}_2(s)e^{-2sT}]}{1 - \bar{\eta}_1(s)\bar{\eta}_2(s)e^{-2sT}}$$

is the *voltage transfer function at the front of the line*.

It is also necessary to know the input impedance of the line. This is given by

$$Z_{in}(s) = \frac{\bar{V}_1(s)}{\bar{I}_1(s)}.$$

At the front of the line we have

$$\bar{I}_1(s) = \bar{I}_\leftarrow(s) + \bar{I}_\rightarrow(s) = \frac{-\bar{V}_\leftarrow(s)}{Z_0} + \frac{\bar{V}_\rightarrow(s)}{Z_0};$$

$$\bar{I}_1(s) = \frac{-\bar{A}(s)\bar{\eta}_2(s)e^{-2sT} + \bar{A}(s)}{1 - \bar{\eta}_1(s)\bar{\eta}_2(s)e^{-2sT}} \frac{\bar{V}_0(s)}{Z_0};$$

$$\bar{I}_1(s) = \frac{\bar{A}(s)(1 - \bar{\eta}_2(s)e^{-2sT})}{1 - \bar{\eta}_1(s)\bar{\eta}_2(s)e^{-2sT}} \frac{\bar{V}_0(s)}{Z_0}.$$

Hence,

$$Z_{in}(s) = \frac{1 + \bar{\eta}_2(s)e^{-2sT}}{1 - \bar{\eta}_2(s)e^{-2sT}} Z_0,$$

which depends on the terminating impedance $\bar{Z}_2(s)$ through $\bar{\eta}_2(s)$.

Let us now evaluate the above relations in cases of particular interest. Consider first a transmission line terminated in a resistance Z_0 and driven by a voltage source with a resistive internal impedance of magnitude Z_0, as shown in Fig. 3-3.

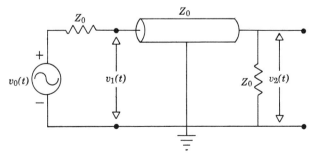

FIGURE 3-3. Transmission line terminated in Z_0 at both ends.

TRANSMISSION AND DELAY LINES

We have

$$\bar{A}(s) = \tfrac{1}{2};$$
$$\bar{\eta}_1(s) = 0;$$
$$\bar{\eta}_2(s) = 0;$$
$$\bar{T}_1(s) = \tfrac{1}{2};$$
$$\bar{T}_2(s) = \tfrac{1}{2} e^{-sT};$$
$$Z_{\text{in}}(s) = Z_0.$$

The response of the line to a voltage step input $v_0(t) = V u(t)$, as shown in Fig. 3-4, is given by

$$\bar{V}_1(s) = \bar{T}_1(s) \bar{V}_0(s);$$
$$\bar{V}_1(s) = \tfrac{1}{2} \bar{V}_0(s);$$
$$v_1(t) = \frac{V}{2} u(t);$$
$$\bar{V}_2(s) = \bar{T}_2(s) \bar{V}_0(s);$$
$$\bar{V}_2(s) = \tfrac{1}{2} e^{-sT} \frac{V}{s};$$
$$v_2(t) = \frac{V}{2} u(t - T).$$

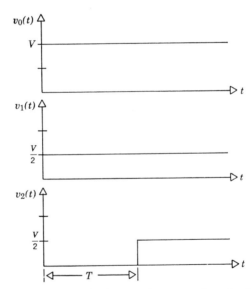

FIGURE 3-4. Voltages at the input, front, and end of a transmission line terminated in Z_0 at both ends.

TERMINATION

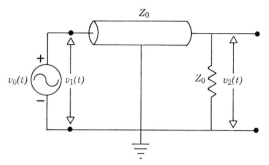

FIGURE 3-5. Transmission line with zero impedance at the front of the line and Z_0 at the end of the line.

The line plus termination "looks like" a resistance of value Z_0 resulting in a voltage division. The voltage at the end of the line is the same as that at the beginning but delayed by the transit time T of the line. We note that the transfer function for a delay circuit is just a constant times e^{-sT}.

Suppose now that the internal impedance of the voltage source is zero, as indicated in Fig. 3-5. We have

$$\bar{A}(s) = 1;$$
$$\bar{\eta}_1(s) = -1;$$
$$\bar{\eta}_2(s) = 0;$$
$$\bar{T}_1(s) = 1;$$
$$\bar{T}_2(s) = e^{-sT};$$
$$Z_{in}(s) = Z_0.$$

We have again a delay circuit, but now there is not the loss of a factor of two in amplitude. This illustrates the desirability of having a low output impedance for a voltage source if one wishes to transmit maximum voltage.

Suppose now we have an open-ended line (or one with $\bar{Z}_2 \gg Z_0$), driven with a voltage source of internal impedance Z_0 (Fig. 3-6). We have

$$\bar{A}(s) = \tfrac{1}{2};$$
$$\bar{\eta}_1(s) = 0;$$
$$\bar{\eta}_2(s) = 1;$$
$$\bar{T}_1(s) = \tfrac{1}{2}(1 + e^{-2sT});$$
$$\bar{T}_2(s) = e^{-sT};$$
$$\bar{Z}_{in}(s) = \frac{1 + e^{-2sT}}{1 - e^{-2sT}} Z_0 = \frac{Z_0}{\tanh sT}.$$

TRANSMISSION AND DELAY LINES

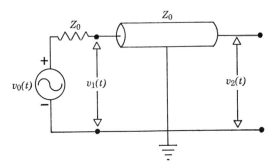

FIGURE 3-6. Transmission line with Z_0 at the front of the line and infinite impedance at the end of the line.

We note that for $|sT| \ll 1$ we may write

$$Z_{in}(s) \simeq \frac{Z_0}{sT} = \frac{1}{s} \frac{\sqrt{\frac{L}{C}}}{l\sqrt{LC}} = \frac{1}{slC},$$

or

$$Z_{in}(s) \simeq \frac{1}{sC_{line}}.$$

Thus, if T is short compared to times of interest (pulse width or half period of a sine wave), the open line, or line terminated in $Z_2 \gg Z_0$, looks like a capacitance equal to the capacitance of the line.

The voltage transfer function $\bar{T}_2(s) = e^{-sT}$ is just the delay-line transfer function; however, the voltage $v_1(t)$ in response to a step input is (Fig. 3-7)

$$v_1(t) = \mathcal{L}^{-1}\left[\frac{V}{2}\frac{(1 + e^{-2sT})}{s}\right] = \frac{V}{2}[u(t) + u(t - 2T)].$$

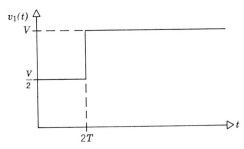

FIGURE 3-7. Voltage at the front of a line terminated in infinite impedance.

TERMINATION

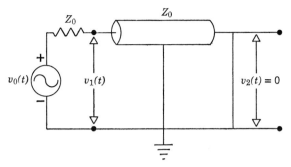

FIGURE 3-8. Transmission line with Z_0 at the front of the line and zero impedance at the end of the line.

Let us now short the transmission line at its output (that is, set $\bar{Z}_2(s) = 0$), as in Fig. 3-8. We have

$$\bar{A}(s) = \tfrac{1}{2};$$
$$\bar{\eta}_1(s) = 0;$$
$$\bar{\eta}_2(s) = -1;$$
$$\bar{T}_1(s) = \tfrac{1}{2}(1 - e^{-2sT});$$
$$\bar{T}_2(s) = 0;$$
$$Z_{\text{in}}(s) = Z_0 \tanh sT.$$

Let us determine $v_1(t)$ in response to a step input $v_0(t) = Vu(t)$. Then

$$\bar{V}_1(s) = \frac{V}{2} \frac{(1 - e^{-s2T})}{s},$$

which is just the Laplace transform of a rectangular pulse of height $V/2$ and width $2T$ (Fig. 3-9). This function of a delay line is used frequently and is

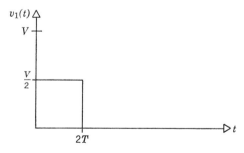

FIGURE 3-9. Voltage at the front of a line terminated in zero impedance (single delay-line clipping).

85

TRANSMISSION AND DELAY LINES

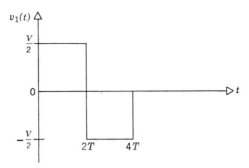

FIGURE 3-10. Double delay-line clipping.

called *single delay-line clipping* or *delay-line pulse shaping*. A short pulse width is necessary when counting random events if "pile-up" is to be kept to a minimum.

It is also of interest to determine the response of the delay-line clip to a rectangular pulse similar to that which might be produced by a preceding delay-line clip of the same type and length. Assume

$$\bar{V}_0(s) = V\frac{(1 - e^{-2sT})}{s},$$

then

$$\bar{V}_1(s) = \tfrac{1}{2}(1 - e^{-2sT})\left[V\frac{(1 - e^{-2sT})}{s}\right]$$

$$= \frac{V}{2}\left[\frac{1 - 2e^{-2sT} + e^{-4sT}}{s}\right];$$

$$v_1(t) = \frac{V}{2}[u(t) - 2u(t - 2T) + u(t - 4T)].$$

This type of pulse shaping is called *double delay-line clipping* (Fig. 3-10) and has two advantages over single delay-line clipping. Since the average voltage level of a single delay-line clipped pulse is nonzero, high counting rates may lead to dc voltage level shifts in following circuits. The double delay-line clipped pulse has zero average value. In addition, it has a zero level *crossover* time $2T$, which is independent of the pulse height and may be used for accurate timing (in coincidence experiments, etc.).

We also note that a shorted delay line looks like an inductance to the preceding circuit if T is short compared to times of interest. For if $|sT| \ll 1$, we have

$$Z_{\text{in}}(s) \simeq Z_0 sT = s\sqrt{\frac{L}{C}}(l\sqrt{LC}) = slL;$$

$$Z_{\text{in}}(s) \simeq sL_{\text{line}}.$$

TERMINATION

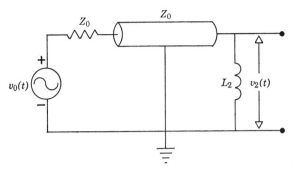

FIGURE 3-11. Transmission line terminated in an inductance.

An illustration will now be given for the case of a reactive termination $Z_2(s)$. We consider the circuit in Fig. 3-11. We have

$$\bar{A}(s) = \tfrac{1}{2};$$

$$\bar{\eta}_1(s) = 0;$$

$$\bar{\eta}_2(s) = \frac{Z_2(s) - Z_0}{Z_2(s) + Z_0} = \frac{sL_2 - Z_0}{sL_2 + Z_0};$$

$$\bar{T}_1(s) = \frac{1}{2}\left[1 + \left(\frac{sL_2 - Z_0}{sL_2 + Z_0}\right)e^{-2sT}\right];$$

$$\bar{T}_2(s) = \frac{s}{s + \dfrac{Z_0}{L_2}} e^{-sT}.$$

Hence, the circuit may be considered a *delayed differentiator* with time constant $\tau = L_2/Z_0$ (Fig. 3-12).

The response to a step input is found to be

$$v_2(t) = V e^{-(t-T)/\tau} u(t - T).$$

An interesting point to note is that there is no reduction in amplitude. The voltage at the front of the line is

$$v_1(t) = \frac{V}{2}[u(t) - u(t - 2T)] + V e^{-(t-2T)/\tau} u(t - 2T).$$

TRANSMISSION AND DELAY LINES

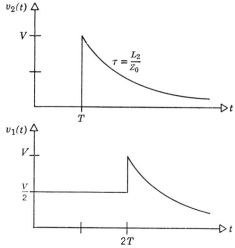

FIGURE 3-12. Voltages at the end and front of a line terminated in an inductance.

The input impedance is

$$Z_{in}(s) = \frac{1 + \left(\frac{sL_2 - Z_0}{sL_2 + Z_0}\right)e^{-2sT}}{1 - \left(\frac{sL_2 - Z_0}{sL_2 + Z_0}\right)e^{-2sT}}$$

$$= \frac{sL_2(1 + e^{-2sT}) + Z_0(1 - e^{-2sT})}{sL_2(1 - e^{-2sT}) + Z_0(1 + e^{2sT})};$$

$$Z_{in}(s) = \left(\frac{Z_0 \tanh sT + sL_2}{Z_0 + sL_2 \tanh sT}\right) Z_0.$$

For $|sT| \ll 1$ we have

$$Z_{in}(s) \simeq sZ_0 T + sL_2;$$

$$Z_{in}(s) \simeq s(L_{\text{line}} + L_2),$$

so the line appears to be two inductors in series.

So far we have considered only the simplest cases such that the line was terminated in the characteristic impedance at either one end or the other, that is, either $\bar{\eta}_1(s) = 0$ or $\bar{\eta}_2(s) = 0$. We may arrive at a rather general interpretation of the situation for finite reflection coefficients at both ends by considering the function

$$\frac{1}{1 - \bar{\eta}_1(s)\bar{\eta}_2(s)e^{-2sT}} = \frac{1}{1 - w(s)}.$$

TERMINATION

If we restrict ourselves to Re $s > 0$, then $|w(s)| < 1$ since $|\bar{\eta}_1(s)| < 1$ and $|\bar{\eta}_2(s)| < 1$. The following power series expansion is then valid:

$$\frac{1}{1 - \bar{\eta}_1\bar{\eta}_2 e^{-2sT}} = 1 + \bar{\eta}_1\bar{\eta}_2 e^{-2sT} + (\bar{\eta}_1\bar{\eta}_2)^2 e^{-4sT} + \cdots.$$

We now evaluate $\bar{V}_1(s)$ and $\bar{V}_2(s)$:

$$\bar{V}_2(s) = \bar{A}(1 + \bar{\eta}_2)e^{-sT}[1 + \bar{\eta}_1\bar{\eta}_2 e^{-2sT} + (\bar{\eta}_1\bar{\eta}_2)^2 e^{-4sT} + \cdots]\bar{V}_0,$$

or *1st received wave* at end of line:

$$\bar{V}_2(s) = \bar{A}\bar{V}_0 e^{-sT} + \bar{\eta}_2 \bar{A}\bar{V}_0 e^{-sT}$$
1st forward wave   1st reflected wave
at end of line   at end of line
at $t = T$ at $t = T$

2nd received wave at end of line:

$$+ \bar{\eta}_1(\bar{\eta}_2 \bar{A}\bar{V}_0)e^{-3sT} \quad + \bar{\eta}_2(\bar{\eta}_1\bar{\eta}_2 \bar{A}\bar{V}_0)e^{-3sT}$$
2nd forward wave   2nd reflected wave
at end of line   at end of line
at $t = 3T$ at $t = 3T$

$$+ \bar{\eta}_1(\bar{\eta}_2\bar{\eta}_1\bar{\eta}_2 \bar{A}\bar{V}_0)e^{-5sT} + \bar{\eta}_2(\bar{\eta}_1\bar{\eta}_2\bar{\eta}_1\bar{\eta}_2 \bar{A}\bar{V}_0)e^{-5sT} + \cdots.$$

At the front of the line we have

1st received wave at front of line:

$$\bar{V}_1(s) = \bar{A}\bar{V}_0 \quad + \bar{\eta}_2(\bar{A}\bar{V}_0)e^{-2sT} \quad + \bar{\eta}_1(\bar{\eta}_2\bar{A}\bar{V}_0)e^{-2sT}$$
1st forward   1st reflected   2nd forward
wave at front   wave   wave at $t = 2T$
of line   arriving
at $t = 0$ at front
                 of line at
                 $t = 2T$

$$+ \bar{\eta}_2(\bar{\eta}_1\bar{\eta}_2 \bar{A}\bar{V}_0)e^{-4sT} + \bar{\eta}_1(\bar{\eta}_2\bar{\eta}_1\bar{\eta}_2 \bar{A}\bar{V}_0)e^{-4sT} + \cdots.$$

We see, then, that if the line is not matched properly, there will be multiple reflections. As an example, let us consider an input wave $v_0(t) = v_0[u(t) - u(t - T/3)]$ and a line terminated with $Z_1(s) = Z_0/2$, $Z_2(s) = 2Z_0$. The voltage $V_2(s)$ is shown in Fig. 3-13.

TRANSMISSION AND DELAY LINES

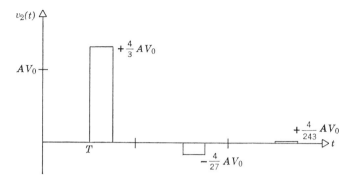

FIGURE 3-13. Voltage at the end of a line with $Z_1 = Z_0/2$ and $Z_2 = 2Z_0$ for an input pulse $v_0(t) = V_0[u(t) - u(t - T/3)]$.

ATTENUATION

The loss-less transmission line serves adequately as a model for a real transmission line except in one respect. In a real transmission line the signal is always somewhat attenuated. We can include this effect in our model if we consider a lossy-leaky line. The attenuation is due to two effects, the resistive loss along the line and the leakage between the conductors. Adding these two effects give us these equations:

$$\frac{\partial v}{\partial z} = -Ri - L\frac{\partial i}{\partial t};$$

$$\frac{\partial i}{\partial z} = -Gv - C\frac{\partial v}{\partial t},$$

where R is the resistance per unit length and G is the conductance per unit length. The Laplace transformed equations, assuming zero initial conditions, are

$$\frac{\partial \bar{V}}{\partial z} = -(R + sL)\bar{I} = -sL'\bar{I};$$

$$\frac{\partial \bar{I}}{\partial z} = -(G + sC)\bar{V} = -sC'\bar{V},$$

where

$$L' = L\left(1 + \frac{R}{sL}\right);$$

$$C' = C\left(1 + \frac{G}{sC}\right).$$

NON-INDUCTIVE TRANSMISSION LINE

The solution of the equations follows in exactly the same way as before with L' and C' replacing L and C. Hence,

$$T = l\sqrt{L'C'} = l\sqrt{LC}\left[\left(1 + \frac{R}{sL}\right)\left(1 + \frac{G}{sC}\right)\right]^{1/2},$$

which becomes

$$T \simeq l\sqrt{LC}\left[1 + \left(\frac{R}{L} + \frac{G}{C}\right)\frac{1}{2s}\right]$$

if $R \ll sL$ and $G \ll sC$. Also,

$$Z_0 = \sqrt{\frac{L'}{C'}} = \sqrt{\frac{L}{C}}\left[\frac{1 + \dfrac{R}{sL}}{1 + \dfrac{G}{sC}}\right]^{1/2},$$

or

$$Z_0 \simeq \sqrt{\frac{L}{C}}\left[1 + \left(\frac{R}{L} - \frac{G}{C}\right)\frac{1}{2s}\right].$$

The forward voltage wave then becomes

$$\bar{V}_\rightarrow(s, z) \simeq \bar{V}(s)e^{-\alpha z}e^{-s\sqrt{LC}\,z},$$

with

$$\alpha = \frac{R}{2}\sqrt{\frac{C}{L}} + \frac{G}{2}\sqrt{\frac{L}{C}}$$

the *attenuation constant* of the line.

NON-INDUCTIVE TRANSMISSION LINE

What would happen if we had a transmission line consisting of resistors and capacitors and only negligible inductance? This situation approximates that of a string of impedance-matched amplifiers, stages each one of which acts as an integrating circuit in the high-frequency approximation. A section of the line is represented symbolically in Fig. 3-14.

The Laplace transform equations, using the dc circuit scheme, are

$$\frac{\partial \bar{V}(s)}{\partial z} = -R\bar{I}(s);$$

$$\frac{\partial \bar{I}(s)}{\partial z} = -(G + sC)\bar{V}(s);$$

and

$$\frac{\partial^2 \bar{V}(s)}{\partial z^2} = R(G + sC)\bar{V}(s).$$

TRANSMISSION AND DELAY LINES

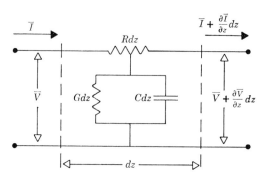

FIGURE 3-14. RC transmission line.

For simplicity, consider a line which is not leaky. The equations simplify to

$$\frac{\partial \bar{V}(s)}{\partial z} = -R\bar{I}(s);$$

$$\frac{\partial \bar{I}(s)}{\partial z} = -sC\bar{V}(s);$$

and

$$\frac{\partial^2 \bar{V}(s)}{\partial z^2} = sRC\bar{V}(s).$$

Using the inverse transform, the equation is seen to be not a wave equation, as before, but a *diffusion equation:*

$$\frac{\partial v}{\partial t} = D \frac{\partial^2 v}{\partial z^2} \quad \text{and likewise} \quad \frac{\partial i}{\partial t} = D \frac{\partial^2 i}{\partial z^2},$$

where $D \, (= 1/RC)$ is the diffusion coefficient, R is the resistance per unit length, and C is the capacitance per unit length. For this reason, such an array of resistors and capacitors is sometimes called a *diffusion line*. The equivalent circuit could be represented, for a finite line, as shown in Fig. 3-15.

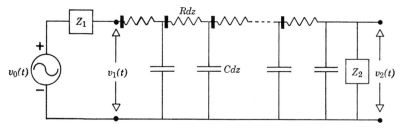

FIGURE 3-15. Schematic of an RC, or diffusion transmission line.

NON-INDUCTIVE TRANSMISSION LINE

The small boxes are ideal impedance-matching networks with infinite input impedance and zero output impedance. Solving the Laplace transformed equations in this case yields solutions for $+z$ and $-z$ as before, but the form becomes

$$\bar{V}(s, z) = \bar{V}_{\leftarrow}(s)e^{\sqrt{s/D}\,z} + \bar{V}_{\rightarrow}(s)e^{-\sqrt{s/D}\,z}.$$

For infinitely long lines, we can reject the first term in $\bar{V}(s, z)$ since it would give us an infinitely large voltage as z approached infinity.

Let us consider the case where $\bar{V}(s)$ is a delta function (*spike*) voltage pulse. Under this assumption, and assuming an infinite line, we can find the inverse transform for the voltage as a function of time and distance along the line. The necessary transform relations* are

$f(t)$	$\bar{F}(s)$
$\dfrac{ae^{-a^2/4t}}{2\sqrt{\pi t^3}}$	$e^{-a\sqrt{s}}$

Thus

$$v(t, z) = \frac{V_0 z e^{-z^2/4Dt}}{2\sqrt{\pi D t^3}}$$

and

$$i(t, z) = \frac{I_0 z e^{-z^2/4Dt}}{2\sqrt{\pi D t^3}}.$$

The subsequent behavior of the voltage pulse is shown in Fig. 3-16. The pulse travels down the diffusion line, constantly broadening, or spreading out, and decreasing in magnitude until finally it dies out as z approaches infinity.

Recalling that

$$\frac{\partial \bar{V}(s)}{\partial z} = -R\bar{I}(s); \qquad \bar{V}(s) = \bar{V}_{\rightarrow}(s)e^{-\sqrt{s/D}\,z};$$

$$\frac{\partial \bar{I}(s)}{\partial z} = -sC\bar{V}(s); \qquad \bar{I}(s) = \bar{I}_{\rightarrow}(s)e^{-\sqrt{s/D}\,z},$$

we find that the relationship between the forward current wave and forward voltage wave becomes

$$\frac{\bar{V}(s)}{\bar{I}(s)} = \frac{\bar{V}_{\rightarrow}(s)}{\bar{I}_{\rightarrow}(s)} = R\sqrt{D/s} = \sqrt{\frac{R}{sC}} \equiv Z_0(s).$$

---

* See any good set of tables as, for example, in the *Handbook of Chemistry and Physics*, Chemical Rubber Publishing Co., Cleveland, Ohio.

TRANSMISSION AND DELAY LINES

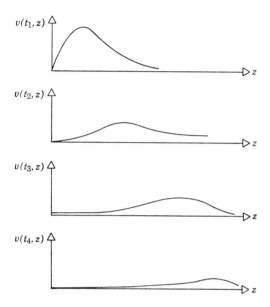

FIGURE 3-16. Time behavior of a pulse in a diffusion line.

Therefore, we now have a frequency-dependent characteristic impedance of the line, which is no surprise since the pulse is obviously being distorted. Other properties of the diffusion line are treated in the problems at the end of the chapter.

RF TRANSMISSION LINES

The response of transmission lines to RF signals leads to some interesting qualitative effects. In the first place, since we are dealing with traveling waves, and have both forward and backward waves in any given line, we have the possibility of obtaining standing waves when proper boundary conditions— that is, terminal impedances—are imposed upon the system. We may examine the detailed behavior of RF lines by using the results we have already obtained for pulses by utilizing the same trick which we developed in Chapter 2; that is, we noted that the solution of any circuitry problem involving zero initial conditions could be obtained for sinusoidal inputs by just replacing the complex variable s in our Laplace transform analysis by $j\omega$. Let us first examine the input impedance to an RF transmission line.

RF TRANSMISSION LINES

We have for the loss-less line,

$$\hat{Z}_{in}(\omega) = \frac{1 + \hat{\eta}_2 e^{-2j\omega T}}{1 - \hat{\eta}_2 e^{-2j\omega T}} Z_0,$$

where

$$\hat{\eta}_2 = \frac{\hat{Z}_2 - Z_0}{\hat{Z}_2 + Z_0}.$$

or since $\omega T = \omega l/v = \beta l$,

$$\hat{Z}_{in} = Z_0 \left(\frac{\hat{Z}_2 + jZ_0 \tan \beta l}{Z_0 + j\hat{Z}_2 \tan \beta l} \right).$$

We now consider two special cases of *short* lines. The input impedance to a shorted line is

$$Z_{in} = jZ_0 \tan \beta l \qquad Z_2 = 0,$$

and for a short line, that is, $\beta l \ll 1$, it becomes

$$\hat{Z}_{in} \simeq j\omega L_{eff},$$

where

$$L_{eff} = \frac{l}{v} Z_0.$$

Thus, a shorted short line appears to the preceding circuit to behave just like an inductor in response to RF signals. An open line, on the other hand, has the following impedance:

$$\hat{Z}_{in} = -jZ_0 \cot \beta l \qquad Z_2 = \infty,$$

which for a short line becomes

$$\hat{Z}_{in} \simeq \frac{1}{j\omega} C_{eff},$$

where

$$C_{eff} = \frac{l}{v} Z_0.$$

A simple interpretation of the above two examples is possible if we recall that

$$v = \frac{1}{\sqrt{LC}}$$

and

$$Z_0 = \sqrt{\frac{L}{C}}.$$

TRANSMISSION AND DELAY LINES

The shorted line looks like an inductor with an inductance equal to the inductance per unit length times the length of line:

$$L_{\text{eff}} = lL.$$

An open line appears to be capacitive with a capacitance equal to the capacitance per unit length times the length of the line:

$$C_{\text{eff}} = lC.$$

In each case we may speak of the "inductance" and the "capacitance" of the line.

We mentioned above that for RF signals we may have standing waves in transmission lines. It is of interest, therefore, to look at the response of particular lengths of lines compared to the wavelength of the exciting sinusoidal source. Let us first look at a *half-wavelength line*. We first note that

$$\beta l = \frac{\omega l}{v} = \frac{2\pi f l}{v} = 2\pi \left(\frac{l}{\lambda}\right).$$

The input impedance becomes

$$\hat{Z}_{\text{in}} = Z_0 \left[\frac{\hat{Z}_2 + jZ_0 \tan 2\pi \left(\frac{l}{\lambda}\right)}{Z_0 + j\hat{Z}_2 \tan 2\pi \left(\frac{l}{\lambda}\right)} \right],$$

and for $l = \lambda/2$ we have, since $\tan \pi = 0$,

$$\hat{Z}_{\text{in}} = \hat{Z}_2.$$

Thus, the effect of the line is completely eliminated. This is true, however, only if we neglect attenuation.

The *quarter-wavelength line* also has some interesting properties. In this case ($l = \lambda/4$) we have, since $\tan \pi/2 = \infty$,

$$\hat{Z}_{\text{in}} = \frac{Z_0^2}{\hat{Z}_2}.$$

The quarter-wavelength line may be used as an impedance matcher since we may obtain a desired impedance $|\hat{Z}_{\text{in}}|$ from a given impedance $|\hat{Z}_2|$ by inserting a quarter-wavelength line in front of $\hat{Z}_2$ with the characteristic impedance

$$Z_0 = \sqrt{|Z_{\text{in}}| |Z_2|}.$$

NMR PROBES

NMR PROBES

The circuit of Fig. 3-17 is important as an NMR (nuclear magnetic resonance) probe. The voltage at the input to the line $\hat{V}_1 = \hat{I}\hat{Z}_{in}$ is monitored; hence, it is useful to look at the input impedance of this line. Using the fact that

$$\hat{Z}_2 = \frac{j\omega L_2}{1 - \omega^2 L_2 C_2},$$

we have

$$\hat{Z}_{in} = jZ_0 \left[\frac{\dfrac{\omega L_2}{1 - \omega^2 L_2 C_2} + Z_0 \tan \dfrac{\omega l}{v}}{Z_0 - \dfrac{\omega L_2}{1 - \omega^2 L_2 C_2} \tan \dfrac{\omega l}{v}} \right].$$

Thus, we will have a series of *resonances* which occur when the denominator becomes zero. (See Problem 3-11 for the case of finite attenuation in the line.)

The condition for resonance is

$$\tan \frac{\omega l}{v} = \frac{Z_0}{\omega L_2}(1 - \omega^2 L_2 C_2).$$

Although this transcendental equation must be solved either graphically or with a computer for ω, we can look at the first few resonances. Figure 3-18 indicates the approximate graphical solution for the case $\omega^2 L_2 C_2 \ll 1$. The first resonance corresponds roughly to the inductance L_2 resonating with the capacitance of the line $C_{eff} = lC$. At low frequencies—that is, long wavelengths—the line is a short line. Since we are driving the line with an ideal

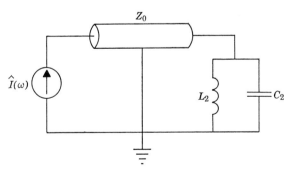

FIGURE 3-17. Schematic of an NMR probe.

TRANSMISSION AND DELAY LINES

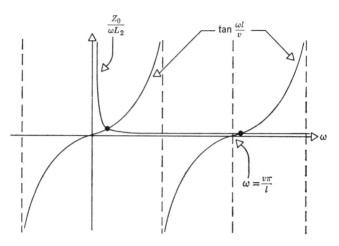

FIGURE 3-18. Graphical solution for an NMR prove for $\omega^2 L_2 C_2 \ll 1$.

current source—that is, infinite impedance—the line, from the point of view of the inductance L_2, looks like a capacitor with capacitance C_{eff}. Using the approximation $\tan \omega l/v \simeq \omega l/v$, it is easily shown that in the vicinity of the first resonance

$$\hat{Z}_{\text{in}} \simeq \frac{j\omega(L_2 + L_{\text{eff}})}{1 - \omega^2 L_2 C_{\text{eff}}}.$$

this corresponds to a parallel tank circuit with inductance L_2 and capacitance C_{eff} if $L_{\text{eff}} \ll L_2$.

As can be seen from Figure 3-18, the second resonance occurs near $l = \lambda/2$, that is, $\omega = (v/l)\pi$. In Fig. 3-19 we illustrate the nature of these resonances for the case of a RG-58/U plus 50 Ω *trombone* (*variable length*) *line* connected to a 1 μh inductance. We have used the following values for v and Z_0:

$$v = 2 \times 10^8 \text{ m/sec};$$

$$Z_0 = 52 \text{ Ω}.$$

The dots in Fig. 3-19 correspond to experimental values while the curves are calculated as indicated in the discussion above. In NMR probe applications, the system is usually operated near the half-wavelength condition in order to minimize the effect of the transmission line.

To see what effect finite attenuation has upon the behavior of RF transmission lines, let us consider the open quarter-wavelength line near its resonant condition $l = \lambda/4$. We first note that the input impedance of a

NMR PROBES

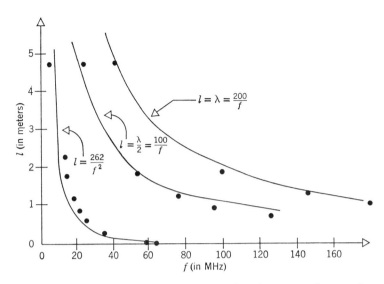

FIGURE 3-19. Comparison between the theory and experiment for an NMR probe.

transmission line with attenuation is

$$\hat{Z}_{\text{in}} = Z_0 \left[\frac{1 + \hat{\eta}_2 e^{-j4\pi(l/\lambda) - 2\alpha l}}{1 - \hat{\eta}_2 e^{-j4\pi(l/\lambda) - 2\alpha l}} \right].$$

Let us excite the line with an RF signal with wavelength

$$\lambda = 4l + \Delta\lambda.$$

Now if $\Delta\lambda/4l \ll 1$ and $2\alpha l \ll 1$, we may write

$$-j\frac{4\pi l}{\lambda} - 2\alpha l \simeq -j\pi\left(1 - \frac{\Delta\lambda}{4l}\right) - 2\alpha l$$

and

$$e^{-j(\pi - \pi(\Delta\lambda/4l)) - 2\alpha l} \simeq -1\left(1 + j\pi\frac{\Delta\lambda}{4l} - 2\alpha l\right).$$

Thus, keeping only first-order terms, we have

$$\hat{Z}_{\text{in}} \simeq Z_0 l\alpha \left[1 + \frac{2j\pi}{4l\alpha}\left(-\frac{\Delta\lambda}{4l}\right)\right],$$

or

$$\hat{Z}_{\text{in}} \simeq Z_0 \alpha l \left[1 + \frac{2j\pi}{\alpha\lambda_0}\left(-\frac{\Delta\lambda}{\lambda_0}\right)\right],$$

99

TRANSMISSION AND DELAY LINES

where
$$\lambda_0 = 4l,$$
but
$$-(\Delta\lambda/\lambda_0) = \Delta\omega/\omega_0.$$
Thus,
$$\hat{Z}_{in} \simeq Z_0 \alpha l \left[1 + j2\left(\frac{\pi}{\alpha\lambda_0}\right)\frac{\Delta\omega}{\omega_0}\right].$$

Comparing this with the results of Problem 2-12 for a series RCL circuit allows us to make the following interpretation:

$$\hat{Z} = R\left[1 + j2Q\frac{\Delta\omega}{\omega_0}\right].$$

The open $\lambda/4$ line looks like a series resonant RCL circuit with resistance

$$R_{eff} = \alpha l Z_0$$

and has a resonant Q given by

$$Q = \frac{\pi}{\lambda_0 \alpha}.$$

Except in special cases the magnitude of the forward and backward waves in an RF transmission line will not be equal. In this case the standing-wave voltage amplitude will vary along the line and there will be a maximum and a minimum value of voltage amplitude. The ratio of maximum amplitude to minimum amplitude may be readily measured and is called the *standing wave ratio*, or SWR.

$$SWR = V_{max}/V_{min}$$

The value of the SWR is that knowledge of it enables us to calculate the magnitude of the reflection coefficient at the end of the line. To see this, we recall that at any point along the line we have a superposition of a backward and a forward wave:

$$\hat{V}(\omega, z) = V_\leftarrow e^{j(\omega/v)z} - V_\rightarrow e^{-j(\omega/v)z}.$$

But from our previous work we found that

$$V_\leftarrow = \hat{\eta}_2 e^{-2j\omega T} V_\rightarrow.$$

Hence,
$$\frac{\hat{V}(\omega, z)}{V_\rightarrow} = |\hat{\eta}_2| e^{j(\delta - 2(\omega/v)l)} e^{j(\omega/v)z} + e^{-j(\omega/v)z},$$
where
$$\hat{\eta}_2 = |\hat{\eta}_2| e^{j\delta}$$

PROBLEMS

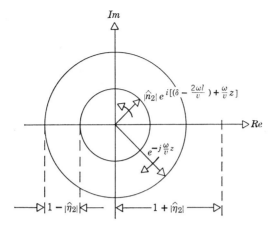

FIGURE 3-20. Voltage phasors in an RF transmission line.

Then at any point along the line the voltage phasor is the superposition of two phasors rotating in opposite directions, as we move down the line along z, with a phase difference. The important point, however, is that their amplitudes differ so that the voltage phasor will be maximum when they add up and minimum when they subtract.

From Fig. 3-20 it can be seen that

$$SWR = \frac{1 + |\hat{\eta}_2|}{1 - |\hat{\eta}_2|}.$$

PROBLEMS

1. Consider a transmission line with the following terminations:

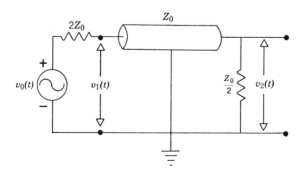

Determine the responses of this line, $v_1(t)$ and $v_2(t)$, to a rectangular pulse:

$$v_0(t) = V[u(t) - u(t - T_0)],$$

TRANSMISSION AND DELAY LINES

where
$$T_0 < T.$$
Sketch the outputs.

2. Calculate the responses, $v_1(t)$ and $v_2(t)$, of the following circuit to a step input.

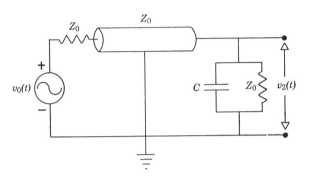

Sketch the outputs.

3. Derive the value of the output impedance of a transmission line. Show that if $\bar{Z}_1 = Z_0$, $\bar{Z}_{\text{out}}(s) = \bar{Z}_2 Z_0/(\bar{Z}_2 + Z_0)$.

4. Consider two transmission lines, each terminated in its characteristic impedance. Suppose that we wish to pass a pulse from one line into the other, using some sort of black box that contains only resistors. We wish, of course, to avoid reflections and excessive attenuation.

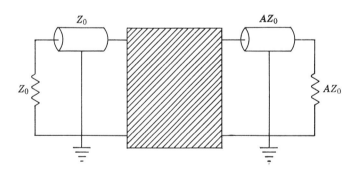

(a) Design a circuit that allows a pulse to pass from one line to the other in either direction without reflections.
(b) Are there any limits to A for this device to work?
(c) What is the attenuation in either direction?
(d) What are the values of the resistors for a 50 Ω-to-125 Ω match?

PROBLEMS

5. In some cases, it is necessary to have a capacitor before a transmission line, often to block high voltage from the line.

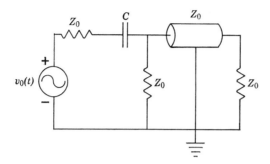

   (a) If all resistors and Z_0 are 50 Ω and the capacitor is 0.1 microfarad, plot the output pulse for a 10^{-6}-second-long unipolar rectangular input pulse with voltage V_0.
   (b) Modify the source impedances to improve the output pulse. Draw the circuit and the resultant pulse.

6. A pulse generator with an output impedance of 50 ohms is used to drive a line rated at 93 ohms. The line is terminated in another 50-ohm impedance which is also purely resistive. For an input pulse shorter than the transit time of the line T, sketch the output at the front end of the line and the back end of the line until the reflected pulses are less than 0.1% of the input pulse. Note: The repetition rate of the pulser is much less than the time scale considered here.

7. Consider a diffusion line of infinite extent. For a step input, show that the voltage at any point and time is

$$v(z, t) = V_0 \left(1 - \mathrm{erf}\left(\frac{z}{2\sqrt{DT}}\right)\right).$$

   What does the voltage look like at any given time along the line? Sketch.

8. Show that the transfer function for a diffusion line of finite extent is

$$\bar{V}_2(L, s) = \bar{T}(s) \bar{V}_0(0, s);$$

$$\bar{T}(s) = \frac{\sqrt{s}\, \bar{Z}_2(s) e^{-\sqrt{s/D}\, L}}{R\sqrt{D} + \sqrt{s}\, \bar{Z}_1(s)}.$$

9. Show that the input impedance for an open half-wavelength line *with attenuation* is given by

$$Z_{\mathrm{in}} = Z_0 \coth \frac{\alpha \lambda}{2}.$$

   Compute an order of magnitude for this value for an RG-58/U line and a frequency $f = 40$ MHz.

TRANSMISSION AND DELAY LINES

10. Show that the Q of a parallel resonant RCL circuit is given by $Q = R/\omega_0 L$.
11. Show that, when attenuation of the transmission line and a parallel resistance are taken into account, the energy stored in the inductance of an NMR probe is given by

$$\langle W_L \rangle = \frac{\frac{1}{4}L}{\left[(1 - \omega^2 L_2 C_2)\cos\frac{\omega l}{v} - \frac{\omega L_2}{Z_0}\sin\frac{\omega l}{v}\right]^2 + \left(\frac{\omega L_2}{R_2}\right)^2\left(\cos^2\frac{\omega l}{v} + 2\frac{R_2 \alpha l}{Z_0}\right)}$$

Experiment 3
Transmission Lines

Equipment:

1. Oscilloscope (including external triggering capability)
2. Decade resistance boxes (2)
3. Pulse generator with external trigger
4. Coaxial cable with total delay arround 5 microseconds

I. General

    A. Set up the circuit shown below.

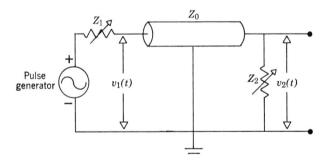

    B. Determine the output impedance of the pulse generator (often 50 ohms) from the manufacturer's specifications. Use the trigger of the generator to trigger the oscilloscope.

II. Measurement of v_0

    A. Remove the transmission line and decade resistors from the output terminal of the pulse generator, and using the cable provided, put the output directly into the oscilloscope. Set the amplitude of the pulse to the maximum value you can measure on the 'scope.

EXPERIMENT 3

B. Note: Since the input impedance of the 'scope is so high, and the source impedance of the pulse generator is so low, the voltage measured in A will be practically v_0.

C. Reconnect the test circuit.

III. Determination of characteristic impedance

A. Set Z_2 to some value $>500 \, \Omega$, Z_1 to zero. (Actually, of course, Z_1 will still be 50 Ω, the internal impedance of the source. The decade resistor is to be set to zero.)

B. While observing the voltage at the input of the transmission line, vary Z_2 until the reflected pulses disappear (as close as you can determine). Note the setting of Z_2. This should equal the characteristic impedance, Z_0.

C. Set Z_2 to some value $>Z_0$, and while observing the voltage at the output of the line, vary Z_1 until the reflected pulses disappear. Note the setting of Z_1. This value (plus 50 Ω) should now equal Z_0.

IV. Measurement of pulse heights and time delays

A. Using the values of Z_1 and Z_2 given below, measure v_{11}, v_{12}, v_{21}, and v_{22} (that is, 1st forward pulse, 2nd forward pulse, 1st received pulse, and 2nd received pulse), and the time delays introduced by the cable. Use the values

$$Z_0 = 50 \, \Omega \qquad Z_2 = 0 \, \Omega$$
$$= \tfrac{1}{2}Z_0 \qquad \quad\; = \tfrac{1}{2}Z_0$$
$$= Z_0 \qquad \quad\;\; = Z_0$$
$$= 2Z_0 \qquad \quad\; = 2Z_0$$

in all possible combinations (that is, set $Z_1 = 50 \, \Omega$, run Z_2 through all four values; $Z_1 = \tfrac{1}{2}Z_0$, run Z_2 through all four values; etc.).

B. Compute the values for v_{11}, v_{12}, v_{21}, and v_{22} from the equations given in class, for each case recorded above, and compare your measured values with the predicted values.

C. Taking the length of the cable to be as marked, calculate the velocity of the pulse along the cable.

V. Delay-line clipper action

Set the pulse length to 10 μsec, $Z_1 = Z_0$, $Z_2 = 0$, and observe and sketch the waveform at the input to the transmission line. Why is the waveform not as predicted by the theory?

TRANSMISSION AND DELAY LINES

VI. Attenuation

From your comparison of computed voltage amplitudes versus measured, try to arrive at a value for the attenuation occurring in the transmission line.

VII. Impedances with reactive elements

A. Connect a decade capacitor in parallel with the decade resistor at Z_2. Set the capacitor to 0.002 μf. Set the decade resistors to equal Z_0.
B. Use 10 μsec pulse length and look at $v_2(T)$ and $v_1(T)$.
C. Measure the rise time and 0–63% time of the pulse occurring at the output and input.
D. Compare with the theory presented in class.

VOLTAGE AMPLIFIER CIRCUITS
TUBES AND FETS

THINK VOLTAGE

At this point, having studied passive elements in various circuit configurations, we are now in a position to consider linear devices which respond *actively* to an input signal, allowing *amplification* to take place. An ideal amplifier is a device that, in response to an input signal, generates an output signal proportional to the input signal. For a voltage amplifier,

$$v_2(t) = Av_1(t),$$

where the constant of proportionality A is the amplification, and can assume any value between zero and (in principle) $\pm$ infinity. If the amplification is negative, the amplifier is said to *invert the signal*. Amplifiers are usually indicated schematically as in Fig. 4-1. The ground line is often omitted.

107

VOLTAGE AMPLIFIERS

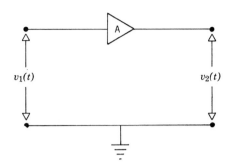

FIGURE 4-1. Ideal voltage amplifier.

We can rewrite this circuit in more detail, using Thevenin's theorem, and exhibit the characteristic impedances of the device explicitly. Using an ideal voltage source, the circuit becomes (in Laplace transform notation) that shown in Fig. 4-2.

For an ideal amplifier, $\bar{T}(s) = A$, the proportionality constant, and the amplifier is equally effective at all frequencies. This is never achieved, naturally, but often the amplification will be approximately constant over a wide frequency range. The amplification in this range is called the *mid-band gain* or *mid-frequency gain*, and the above approximation will be applicable.

As before,* we define $\bar{V}_{\text{open}}(s)$ as the voltage at $\bar{V}_2(s)$ with $\bar{Z}_2(s) = \infty$, and $\bar{I}_{\text{short}}(s)$ as the current that flows through the output circuit when $\bar{Z}_2(s) = 0$, and find

$$\bar{V}_{\text{open}}(s) = \bar{T}(s)\bar{V}_1(s) \qquad \bar{Z}_2(s) = \infty$$

and

$$\bar{I}_{\text{short}}(s) = \bar{T}(s)\bar{V}_1(s)/\bar{Z}_0(s) \qquad \bar{Z}_2(s) = 0,$$

so that

$$\bar{Z}_0(s) = \text{the output impedance} = \bar{V}_{\text{open}}(s)/\bar{I}_{\text{short}}(s).$$

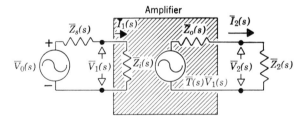

FIGURE 4-2. Thevenin's theorem equivalent circuit for an amplifier.

* See Appendix to Chapter 1.

THINK VOLTAGE

We could equally well have used an ideal current source and Norton's theorem for this analysis, defining a current transfer function $\bar{\beta}(s)$ in a manner analogous to the voltage transfer function $\bar{T}(s)$.

The choice of which of the equivalent analyses to use depends largely on the input impedance of the real devices that constitute the amplifier circuit. If the input impedance Z_i is very high, 10^9 ohms or more, as it is in vacuum tubes and field-effect transistors (FETs), it will be difficult to measure an input current, while voltage measurements are easy, for

$$\bar{V}_1(s) = \bar{I}_1(s)Z_i(s) = \frac{\bar{V}_0(s)Z_i(s)}{Z_i(s) + Z_s(s)} \simeq \bar{V}_0(s)$$

for reasonable values of $Z_s(s)$.

Conversely, if the input impedance is small, the input voltage $\bar{V}_1(s)$ can be significantly different from $\bar{V}_0(s)$. It is often more convenient to use the input current $\bar{I}_1(s)$ as the parameter of interest in these cases, since it can be relatively large and easy to determine accurately. Transistors are a prime example of a device with low input impedance. In intermediate cases (certain configurations of transistors, for example), either analysis or a combination of both analyses can be used, depending on details of the device and its planned usage in the circuit. A preference for one type of analysis over the other may occur if one of the parameters, voltage or current, is more closely connected to the physical principles upon which the operation of the device is based and thus, perhaps easier to calculate.

Therefore, it is natural to separate the analysis into two parts based on the value of the input impedance of the devices. The first considers devices with *high input impedances*, in which the analysis is based on *voltage*, the input current flow entering the analysis only when the frequency becomes so high that capacitive effects become important. The second considers devices with *low input impedances*, in which the analysis is based on *current*. Models suitable for each of these cases will be derived separately, although they will have much in common. We will not spend a lot of time on the physical principles of actual active elements for reasons that are expressed more fully in the preface, namely, that we could not possibly do justice to the intricacies of the physics without expanding the text to unwieldy proportions, that many good texts on the physical principles of electronic devices are presently available, and that we have no doubt that new devices will be constantly appearing on the market which would render obsolete much of the discussion. We will discuss selected devices that are useful in developing techniques which can be used for a wide array of present (and future) active elements, indicating differences that affect performance of the element and modify the approximations which we will make.

VOLTAGE AMPLIFIERS

HIGH-INPUT-IMPEDANCE DEVICES

High-input-impedance devices are characterized by input currents that are usually many orders of magnitude smaller than the currents that they control. One such high-input-impedance device is the vacuum tube.

The vacuum triode is the simplest tube capable of amplification, so we shall construct first a model for the triode which is applicable for *small signals*, for which the tube acts as a *linear device*. The triode consists essentially of three elements in an evacuated glass tube. Two of the elements are the same as those in a vacuum diode, the *plate* and the *cathode*. The latter is constructed of a material which easily "boils" off electrons when heated, thus creating an electron cloud, or space charge, in the vicinity of the cathode. When a positive voltage is applied between the plate and the cathode, the electrons are attracted by the positive plate, giving rise to a flow of *conventional current* from plate to cathode. If the plate voltage is made negative with respect to the grid, the electrons will be repelled by the negative plate, but very little current will flow since the plate is not constructed so that it will give off electrons easily. One then has a thermionic diode. If a third element, consisting of a mesh of wines of small cross section, is placed near the cathode, a small voltage between this *grid* and the cathode can affect appreciably the flow of plate current (Fig. 4-3). If the dc voltage placed on the grid, the grid bias, is such that the grid becomes more negative with respect to the cathode, electrons are repelled from the region of the grid, and the current through the tube decreases for a fixed value of the plate voltage. One of the families of curves that describes the operation of the triode is given (Fig. 4-4) by the dependence of the plate current I_p on the plate voltage V_p for a series of values of the grid bias.

Figure 4-3 defines the directions of the plate current I_p, grid current I_g and cathode current I_K. Kirchhoff's first law requires that

$$I_K = I_p + I_g.$$

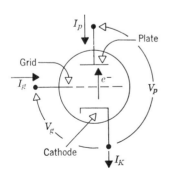

FIGURE 4-3. Schematic of a vacuum triode.

HIGH-INPUT-IMPEDANCE DEVICES

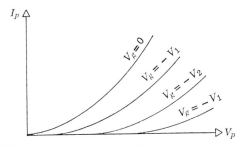

FIGURE 4-4. Plate current versus plate voltage at various values of grid bias for a vacuum triode.

The plate voltage V_p and grid voltage V_g are measured with respect to the cathode and are positive if the numerical values of V_p and V_g are positive.

A solid-state analog to the triode is the field-effect transistor (FET). It consists of a bar of semiconductor, usually silicon or germanium, onto which are fixed two ohmic contacts (metal contacts which merely complete the electrical circuit, so that the whole device acts as a resistor). The contact at one end of the bar is called the *source*, and at the other end, the *drain*. The semiconducting bar is *doped* with acceptors (which makes the bar a *p*-type material with *holes* carrying the current) or donors (which makes the bar an *n*-type material with *electrons* carrying the current). These conducting holes or electrons are the *majority* carriers in the *p*-type or *n*-type semiconductor, respectively, with the *minority* carriers being the much smaller number of the opposite-type carrier.

In a vacuum triode, the majority carriers are the electrons, and the minority carriers are the positive ions that occur due to the small amount of residual gas in the tube. Both contribute to the net current, since they move in opposite directions in the electric field of the tube.

If the semiconducting bar is doped to be *p*-type material, the resulting FET will be called a *p-channel FET*, and for *n*-type material, an *n-channel FET*. The elements which control the current flow from the source to the drain, the analog to the grid of a triode, are called the *gates*. Two types of gate are common: the first consists of two *junction diodes* formed into either side of the semiconducting bar (JFET); the second consists of two insulating layers applied to the surface of the bar (IGFET). When this layer is formed of metal oxide, the resulting device is called a metal-oxide semiconductor (MOS), and is often designated by the term MOS FET. When one uses the term FET, a JFET is usually the type referred to. An *n*-channel FET is illustrated in Fig. 4-5.

VOLTAGE AMPLIFIERS

The biasing of an *n*-channel FET is identical to that of a triode, the drain very positive and the gate slightly negative with respect to the source. A *p*-channel FET reverses both polarities. In either type of FET, applying a bias to the gate (the polarities result in the junction being back biased) removes current carriers from the region of the gate, constricting the current flow in the channel and increasing the effective resistance of the device. For

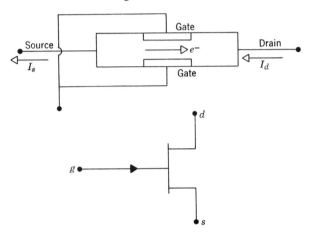

FIGURE 4-5. Schematic of an FET.

small voltages applied between the source and the drain, the current in the device increases linearly, since it simply acts as a resistor.

As the voltage becomes higher, the gates, which are held at a fixed voltage by some external bias (which can be zero), become relatively more highly biased, since the voltage of the channel between the gates is set by the ohmic drop along the bar from the source to the drain. This higher gate bias increases the effective resistance of the bar, and the current flow no longer rises linearly but tends to flatten out, giving a plateau where the drain current I_d is almost a constant. Since the gate is back biased, the current flow through the gate is very low, much lower than I_d in cases of interest. This value of I_d on the plateau defines *pinch-off*, and for each bias on the gates, pinch-off occurs at a different value. At a high enough value of the voltage V_d between the source and the drain, the device breaks down and current suddenly rises. One family of curves that describes the behavior of an *n*-channel FET is given schematically in Fig. 4-6.

The key to the analysis of devices with high input impedances is that they *all* possess similar dependence on the *voltage* of the controlling element (grid, gate, etc.), so that it is convenient to write the current through the

HIGH-INPUT-IMPEDANCE DEVICES

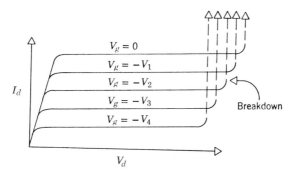

FIGURE 4-6. Drain current versus drain voltage at various values of gate bias for an FET.

device in terms of the voltages on the controlling element and the voltage across the device. For a triode,

$$I_p = I_p(V_g, V_p).$$

For an FET,

$$I_d = I_d(V_g, V_d).$$

Is this the only possible form that we can choose under our assumptions? There are six possible variables, the three currents and the three voltages. One voltage is eliminated since only the *relative* voltage across the device can influence its behavior, and two voltages are measured with respect to the third. Kirchhoff's first law eliminates one of the currents. We *assume* that the grid (or gate) current is zero, which eliminates another. Finally, we have a dependent variable. Therefore, the dependent variable must be a function of only two independent variables, although the choice of which variables will be independent or dependent is still arbitrary. However, one of the independent variables should be the grid (or gate) voltage, since this governs the operation of the device. In other situations, we will argue for other choices of the dependent and independent variables.

The important point is that the formalism applies to *any* high-input-impedance, three-terminal device; tubes, FETs, etc., including devices as yet unknown. We will use the terminology of tubes in the analysis, but at any point, one can substitute gate for grid, drain for plate, and source for cathode, and be talking about an FET. Differences will arise in discussing capacitive effects, which are easier to formulate in the case of tubes, since the capacitances are constant and not a function of the voltages.

The functional dependence indicated by $I_p = I_p(V_g, V_p)$ is, in general, quite complicated and certainly nonlinear. The currents we are considering

VOLTAGE AMPLIFIERS

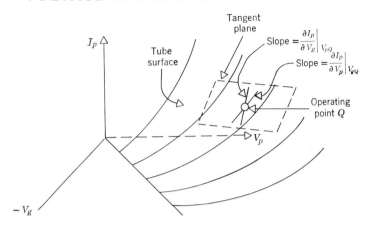

FIGURE 4-7. Tube surface for a vacuum triode.

here might be called *transfer currents* since they represent the time rate of change of charge actually transferred from one element to the other. To have a useful model for reasonably fast pulses, we must later consider the capacitive effects and include *charging currents*. The above equation is a function of two variables and represents a surface, the *tube surface*, in a three-dimensional rectangular space of I_p, V_p, and V_g (Fig. 4-7).

This three-dimensional information is very easily reduced to two dimensions (so that one may construct a catalogue of tube characteristics) by the following technique: to obtain the *plate characteristics*, we select a value of $-V_g$, then construct a plane perpendicular to the $-V_g$ axis at this value of grid voltage. This plane will intersect the tube surface yielding a curve in space. We then translate or project this curve into the (I_p, V_p) plane and label or parameterize the curve by the chosen value of grid voltage. Choosing several values of grid voltage and applying the above techniques gives us the family of curves shown in Fig. 4-8. We might note that this was

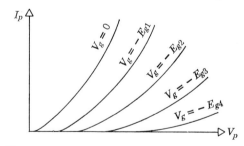

FIGURE 4-8. Plate current versus plate voltage at various values of the grid bias for a vacuum triode.

LINEAR OR SMALL-SIGNAL ANALYSIS

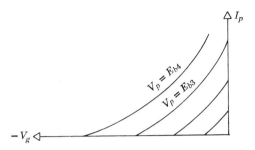

FIGURE 4-9. Plate current versus grid voltage for various values of plate voltage for a vacuum triode.

the projection previously used to obtain the family of curves characteristic of a triode and an FET.

We may also take projections onto the $(-V_g, I_p)$ plane, obtaining the *transfer characteristics*, as shown in Fig. 4-9. Here the various curves are parameterized by the selected values of plate voltage. In actual practice, it is easy to obtain the plate and transfer characteristics experimentally so that one need not construct the tube surface. The above analysis is, of course, just what one does when treating PVT surfaces and curves in thermodynamics.

LINEAR, OR SMALL-SIGNAL, ANALYSIS

If a linear amplifying device is desired, a fairly flat portion of the tube surface is found and an *operating point Q* selected in the center of this region. We then expand the plate current in a double Taylor series about the operating, or quiescent, point Q:

$$I_p(V_g, V_p) = I_{pQ}(V_{gQ}, V_{pQ}) + \frac{\partial I_p}{\partial V_g}\bigg)_{V_{pQ}} \Delta V_g + \frac{\partial I_p}{\partial V_p}\bigg)_{V_{gQ}} \Delta V_p + \cdots .$$

Neglecting all nonlinear terms, we may write this equation as follows:

$$\Delta I_p = g_m \Delta V_g + \frac{1}{r_p} \Delta V_p.$$

This equation is called the *triode equation* and is the equation for the plane tangent to the tube surface at the operating point. For small deviations from the quiescent values (small signals), we effectively substitute the tangent plane for the tube surface. The tangent plane is uniquely determined by the

VOLTAGE AMPLIFIERS

operating point and the slopes of the tangent lines. These slopes are dimensionally conductances and now completely represent the effect of the tube. They are given special names.

$$g_m = \frac{\partial I_p}{\partial V_g}\bigg)_{V_{pQ}}$$

is called the *transconductance* of the tube.

$$\frac{1}{r_p} = \frac{\partial I_p}{\partial V_p}\bigg)_{V_{gQ}}$$

is the reciprocal of the *plate resistance* of the tube.

A consideration of the grid current is much simpler. If the grid is operated at a negative potential with respect to the cathode, we may assume $I_g = 0$. If it is operated at a positive potential with respect to the cathode, there will be grid current flowing, since the grid will now attract electrons. The *tube curve* for the grid-cathode "diode" is similar to an ordinary thermionic diode curve (Fig. 4-10), although the grid is physically very small and is not designed to accept electrons. From the construction of a triode it is found experimentally that the grid current is approximately independent of plate voltage; hence, the grid current depends only upon the grid voltage:

$$I_g = I_g(V_g).$$

Expanding this function about the quiescent grid voltage gives us

$$I_g(V_g) = I_{gQ}(V_{gQ}) + \frac{\partial I_g}{\partial V_g}\bigg)_{V_{gQ}} \Delta V_g + \cdots$$

or

$$\Delta I_g = \frac{1}{r_g} \Delta V_g,$$

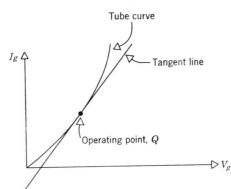

FIGURE 4-10. Definition of the grid resistance of a tube.

LINEAR OR SMALL-SIGNAL ANALYSIS

where

$$\frac{1}{r_g} = \left.\frac{\partial I_g}{\partial V_g}\right)_{V_{gQ}}$$

is the reciprocal of the *grid resistance* of the tube. The above equation is the equation of a straight line tangent to the tube curve at the operating point Q. The slope of the line is just $1/r_g$. For small signals we may approximate the tube curve by the tangent line.

FIGURE 4-11. Triode amplifier.

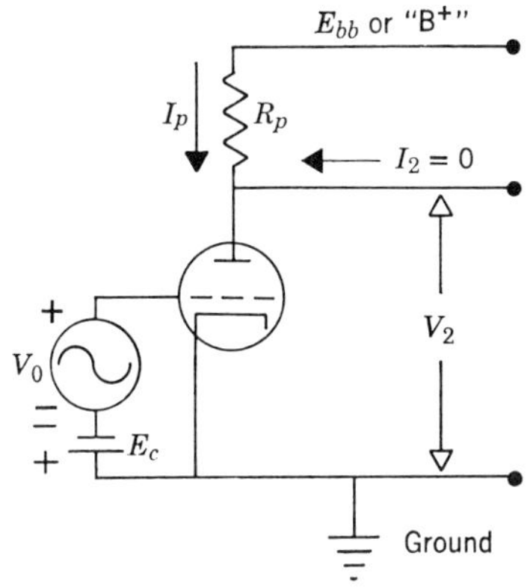

We now indicate a method of setting up the dc, or quiescent, condition for a single-stage triode amplifier. As indicated above, one first decides upon the general region of the operating point by looking for a flat portion of the tube surface. This may be done by looking at the plate and transfer characteristics.

Let us consider the circuit diagram of Fig. 4-11. E_{bb}, or "B-plus," is the positive supply voltage, E_c is the *grid bias* voltage (arranged so that the grid is at a negative voltage with respect to the cathode), and R_p is the plate resistor which is necessary if we are to operate at our desired quiescent point. V_0 is the input signal voltage. Applying Kirchhoff's second law to the plate and grid loops gives us

$$E_{bb} - I_p R_p - V_p = 0;$$
$$-E_c + V_0 - V_g = 0.$$

We wish to solve these equations for the quiescent values I_{pQ}, V_{pQ}, and V_{gQ}. Since V_0 is the signal voltage, its quiescent value is zero; thus, the second equation gives us immediately V_{gQ}:

$$V_{gQ} = -E_c.$$

VOLTAGE AMPLIFIERS

The first equation is linear in I_p and V_p and is thus the equation of a line, the *load line*, in the plane of the plate characteristics:

$$I_p = -\frac{1}{R_p} V_p + \frac{E_{bb}}{R_p}.$$

This equation and the tube equation

$$I_p = I_p(-E_c, V_p)$$

must be solved graphically for the quiescent values of I_{pQ} and V_{pQ} (Fig. 4-12).

In determining the operating point, one must be careful not to exceed the power dissipation limits of the tube, that is, the product $I_p V_p$ must be less than the constant H which determines the *power dissipation hyperbola*:

$$I_p V_p = H.$$

We now compute the *dc gain* of the amplifier stage, which is applicable for slowly varying signals, that is, we ignore capacitors and charging currents. Applying Kirchhoff's second law to the output loop yields

$$V_p - V_2 = 0.$$

This equation, along with those found from the plate loop and grid loop, may be written in terms of the quiescent values and the deviations from quiescent values, or signals. We make the following substitutions:

$$I_p = I_{pQ} + \Delta I_p;$$
$$V_p = V_{pQ} + \Delta V_p;$$
$$V_g = V_{gQ} + \Delta V_g;$$
$$V_0 = 0 + \Delta V_0;$$
$$V_2 = V_{2Q} + \Delta V_2.$$

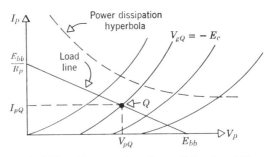

FIGURE 4-12. Determination of the load line.

LINEAR OR SMALL-SIGNAL ANALYSIS

The quiescent terms drop out, giving us
$$\Delta V_p = -\Delta I_p R_p;$$
$$\Delta V_g = \Delta V_0;$$
$$\Delta V_2 = \Delta V_p.$$

We wish to solve for the gain
$$A = \frac{\Delta V_2}{\Delta V_0} = \frac{\Delta V_p}{\Delta V_g}.$$

Eliminating ΔI_p from the first of the above three equations and the triode equation gives us
$$A = -g_m r_{\parallel},$$
where
$$\frac{1}{r_{\parallel}} = \frac{1}{r_p} + \frac{1}{R_p}.$$

We note that the negative sign indicates a reversal of phase. When the grid goes down, the plate goes up.

Once the operating point has been determined, one may graphically find the tube parameters by using
$$g_m \simeq \left.\frac{\Delta I_p}{\Delta V_g}\right|_{V_{pQ}}$$
and
$$\frac{1}{r_p} \simeq \left.\frac{\Delta I_p}{\Delta V_p}\right|_{V_{gQ}}.$$

It is also useful to introduce another tube parameter called the *amplification factor* or μ of the tube, since it is a direct measure of the voltage gain possibilities of the tube. We consider the triode equation for constant plate current, that is, $\Delta I_p = 0$. Hence,
$$g_m \Delta V_g + \frac{1}{r_p}\Delta V_p = 0,$$
or
$$\left.\frac{\Delta V_p}{\Delta V_g}\right|_{I_{pQ}} = -g_m r_p.$$

We define
$$\mu = -\left.\frac{\Delta V_p}{\Delta V_g}\right|_{I_{pQ}} = g_m r_p.$$

The dc gain may then be written in the form
$$A = -\frac{R_p}{R_p + r_p}\mu.$$

VOLTAGE AMPLIFIERS

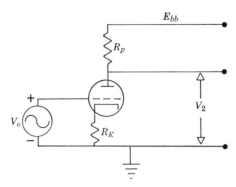

FIGURE 4-13. A self-biasing triode amplifier.

It is possible to eliminate the battery E_c by including a resistor in the cathode-to-ground circuit (Fig. 4-13). The grid is operated at ground potential, which is negative with respect to the cathode. This arrangement is termed *self-biasing*. Determining the operating point is a little more complicated now since we must solve graphically the following equation by iteration for V_{gQ}, I_{pQ}, and V_{pQ}:

$$I_p = -\frac{1}{R_p + R_K} V_p + \frac{E_{bb}}{R_p + R_K} ;$$

$$I_p = I_p(-I_p R_K, V_p).$$

We may repeat the above analysis for the dc gain of the stage to find

$$A = \frac{-g_m r_\|}{1 + \frac{(1+\mu)R_K}{r_p + R_p}} .$$

The dc gain has thus been decreased by the addition of a cathode-biasing resistance. One may eliminate this effect for fast signals and still have self-biasing by placing a large capacitance C_K across R_K, placing the cathode at *ac ground*. This has the effect of holding the cathode voltage V_K relatively constant for fast signals.

TIME RESPONSE OF TRIODE AND FET CIRCUITS

If we wish to work with rapidly varying signals, we must consider the effects of the capacitance that is always present between the elements of a tube or the junctions of a solid-state device. For the case of tubes, the situation is relatively uncomplicated since the *interelectrode capacitances* are fixed by the

TRIODE AND FET TIME RESPONSE

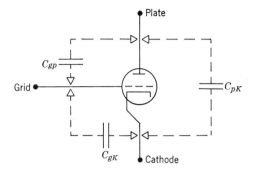

FIGURE 4-14. Interelectrode capacitance in a vacuum triode.

construction of the elements of the tube and are not a function of voltage or current. These capacitances are supplied by the tube manufacturer, and are indicated schematically in Fig. 4-14.

In the case of an FET, we have to first consider the fact that there are two gates (instead of one grid) and, in fact, the interelement capacitances of the two gates will usually be significantly different. Secondly, the gates of a JFET are *pn* junctions, and the capacitance of a *pn* junction becomes smaller as the applied voltage becomes larger, so that an exact treatment of the capacitive effects of an FET would demand that the capacitances be treated as functions of the input voltage. However, for a linear analysis, the input voltage will be held within narrow limits, and the gate biasing is also fairly well fixed by the requirement of maximum linearity, so we will treat the interelement capacitances as constants.

The capacitances associated with an FET are indicated schematically in Fig. 4-15. Since the gate capacitances are in parallel, we will define a total gate-to-source capacitance $C_{gs} = C_{g_1 s} + C_{g_2 s}$. It might be noted at this time that this capacitance is usually quite low, often below 1 pf. Using this combined gate-to-source capacitance (and the similar gate-to-drain capacitance),

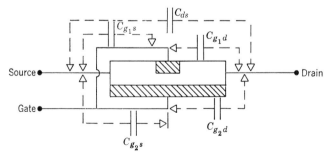

FIGURE 4-15. Interelement capacitance in an FET.

VOLTAGE AMPLIFIERS

we can proceed with an analysis equally valid for FETs and triodes. We will continue to use the terminology of the triode.

The charge stored on the plate and that stored on the grid are given by

$$Q_p^s = C_{pK}V_p + C_{gp}(V_p - V_g);$$
$$Q_g^s = C_{gK}V_g - C_{gp}(V_p - V_g).$$

The total charge on the plate at a given instant is just the sum of the stored charge and the transferred charge:

$$Q_p^{\text{TOT}} = Q_p^s + Q_p^T.$$

The total current is then

$$I_p^{\text{TOT}} = I_p^s + I_p^T,$$

where I_p^T is the transfer current we have considered before. For quiescent conditions we have $Q_{pQ}^s = $ constant. Thus,

$$Q_p^s = Q_{pQ}^s + \Delta Q_p^s = Q_{pQ}^s + q_p^s,$$

where we now return to our usual convention that an increment Δ of a quantity (that is, a signal) will be designated by a small letter. The charging current is given by

$$I_p^s = \frac{dQ_p^s}{dt} = \frac{dq_p^s}{dt} = i_p^s.$$

Using the linear approximation for the transfer current, we have the following for the total current:

$$I_p^{\text{TOT}} = I_{pQ}^T + i_p^T + i_p^s,$$

or

$$I_p^{\text{TOT}} - I_{pQ}^T = \Delta I_p^{\text{TOT}} = i_p^{\text{TOT}} = i_p^T + i_p^s.$$

If we now substitute

$$V_p = V_{pQ} + v_p;$$
$$V_g = V_{gQ} + v_g$$

into the charge equations, the quiescent terms drop out, giving us

$$q_p^s = (C_{pK} + C_{gp})v_p - C_{gp}v_g;$$
$$q_g^s = -C_{gp}v_p + (C_{gK} + C_{gp})v_g.$$

The charging currents are then

$$i_p^s = (C_{pK} + C_{gp})\frac{dv_p}{dt} - C_{gp}\frac{dv_g}{dt};$$

$$i_g^s = -C_{gp}\frac{dv_p}{dt} + (C_{gK} + C_{gp})\frac{dv_g}{dt}.$$

TRIODE AND FET TIME RESPONSE

We add to these the transfer currents

$$i_p^T = g_m v_g + \frac{1}{r_p} v_p;$$

$$i_g^T = \frac{1}{r_g} v_g$$

to give the total currents

$$i_p = g_m v_g + \frac{1}{r_p} v_p - C_{gp} \frac{dv_g}{dt} + (C_{pK} + C_{gp}) \frac{dv_p}{dt};$$

$$i_g = \frac{1}{r_g} v_g + (C_{gK} + C_{gp}) \frac{dv_g}{dt} - C_{gp} \frac{dv_p}{dt}.$$

These equations now describe the tube and may also be called the *triode equations*.

We now Laplace transform the triode equations:

$$\bar{I}_p = \left[\frac{1}{r_p} + s(C_{pK} + C_{gp})\right] \bar{V}_p + (g_m - sC_{gp}) \bar{V}_g;$$

$$\bar{I}_g = -sC_{gp} \bar{V}_p + \left[\frac{1}{r_g} + s(C_{jk} + C_{gp})\right] \bar{V}_g,$$

where the initial condition terms are zero since we are dealing with deviations from equilibrium values and we assume the circuit to be quiescent before a signal is applied. The first term of the second equation represents a feedback effect. There is a charging current contribution to the grid current which depends upon the plate voltage.

A very important point may be brought out about active devices by writing the above triode equations in matrix form:

$$\begin{pmatrix} \bar{I}_p \\ \bar{I}_g \end{pmatrix} = \begin{pmatrix} \bar{a}_{11} & \bar{a}_{12} \\ \bar{a}_{21} & \bar{a}_{22} \end{pmatrix} \begin{pmatrix} \bar{V}_p \\ \bar{V}_g \end{pmatrix},$$

where

$$\bar{a}_{11} = \frac{1}{r_p} + s(C_{pK} + C_{gp});$$

$$\bar{a}_{12} = g_m - sC_{gp};$$

$$\bar{a}_{21} = -sC_{gp};$$

$$\bar{a}_{22} = \frac{1}{r_g} + s(C_{gK} + C_{gp}).$$

The active device may thus be represented by an *admittance matrix* with the matrix elements given above.

VOLTAGE AMPLIFIERS

The triode equations may be simplified in the following approximate way. We assume the existence of the relation

$$\bar{V}_p = \bar{T}\bar{V}_g$$

and write for the triode equations

$$\bar{I}_p = g_m \bar{V}_g + \left(\frac{1}{r_p} + sC_o\right)\bar{V}_p;$$

$$\bar{I}_g = \left(\frac{1}{r_g} + sC_i\right)\bar{V}_g,$$

where

$$C_i = C_{gK} + C_{gp}(1 - \bar{T});$$

$$C_o = C_{pK} + C_{gp}\left(1 - \frac{1}{\bar{T}}\right).$$

To justify the above assumption, let us work out the relation between $\bar{V}_p$ and $\bar{V}_g$ for the simplest situation shown in Fig. 4-11. The ouput of the active device will always be connected to some load impedance, in this case $\bar{Z}_L(s) = R_p$. Thus, in addition to the triode (or active device) equations

$$\bar{I}_p = \bar{a}_{11}\bar{V}_p + \bar{a}_{12}\bar{V}_g;$$

$$\bar{I}_g = \bar{a}_{21}\bar{V}_p + \bar{a}_{22}\bar{V}_g,$$

we have also an external equation

$$\bar{I}_p = \frac{-\bar{V}_p}{\bar{R}_p}.$$

Here we have assumed that the load impedance R_L is very large. Substituting this into the first of the triode equations above and solving for the ratio $\bar{V}_p/\bar{V}_g$ gives us

$$\bar{T}(s) = \frac{\bar{V}_p}{\bar{V}_g} = -\frac{\bar{a}_{12}}{\bar{a}_{11} + \dfrac{1}{R_p}}.$$

Recalling the definition of the admittance matrix elements, we have

$$\bar{T}(s) = -\frac{g_m - sC_{gp}}{\dfrac{1}{R_p} + \dfrac{1}{r_g} + s(C_{pK} + C_{gp})}.$$

TRIODE AND FET TIME RESPONSE

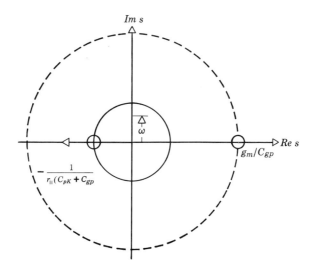

FIGURE 4-16. Triode transfer function.

for the voltage transfer function. We may write this in the following form:

$$\bar{T}(s) = -g_m r_\| \left[\frac{1 - s\dfrac{C_{gp}}{g_m}}{1 + s r_\|(C_{pK} + C_{gp})} \right]$$

where $1/r_\| = 1/R_p + 1/r_p$ as before. The triode transfer function has a pole at $-1/r_\|(C_{pK} + C_{gp})$ and a zero at g_m/C_{gp} (Fig. 4-16).

We note that the pole may be moved out along the negative real axis from $-1/r_p(C_{pK} + C_{gp})$ to $-1/R_p(C_{pK} + C_{gp})$, which could theoretically go all the way to $-\infty$ as R_p is made smaller and smaller. However, this would be at the sacrifice of gain, as we shall see later.

If we consider the above situation with $r_\|(C_{pK} + C_{gp}) > C_{gp}/g_m$, then three different regions may be discerned for the voltage transfer function.

We obtain the magnitude of the frequency dependence of the voltage transfer function by using the transformation $s \to j\omega$ and writing the result in complex polar form:

$$|T(j\omega)| = g_m r_\| \left[\frac{1 + \omega^2 \left(\dfrac{C_{gp}}{g_m}\right)^2}{1 + \omega^2 r_\|^2 (C_{pK} + C_{gp})^2} \right]^{1/2}.$$

Thus, for low frequencies, that is, $\omega < 1/r_\|(C_{pK} + C_{gp})$, we have

$$|T(j\omega)| = g_m r_\| = |A|$$

125

VOLTAGE AMPLIFIERS

for the constant gain of the amplifier. We note that the signal is inverted since the phase in this case is $e^{j\pi} = -1$. For higher frequencies, that is, $1/r_\|(C_{pK} + C_{gp}) < \omega < g_m/C_{gp}$, we have

$$|T(j\omega)| = \frac{g_m r_\|}{\sqrt{1 + \omega^2 r_\|^2 (C_{pK} + C_{gp})^2}},$$

or, correspondingly,

$$\bar{T}(s) = \frac{-g_m r_\|}{1 + s r_\|(C_{pK} + C_{gp})},$$

or

$$\bar{T}(s) = A\left(\frac{1}{1 + s\tau}\right).$$

This looks like an inverter integrating circuit with gain. It is the single-pole approximation for the voltage transfer function of the active device. For very high frequencies, that is, $\omega > g_m/c_{gp}$, we have

$$|T(j\omega)| = \frac{C_{gp}}{C_{pK} + C_{gp}},$$

and the device looks like a passive capacitor voltage divider.

Returning again to the new triode equations, we see that over the useful frequency range of the amplifier we may take

$$\bar{T} \simeq -|A| = -g_m r_\|$$

as a useful approximation in the input and output capacitance equations and assume $|A| \gg 1$. Then

$$C_i \simeq C_{gK} + C_{gp}(1 + |A|);$$

$$C_o \simeq C_{gK} + C_{gp}.$$

The grid has now been effectively decoupled from the plate so that the second of the new triode equations may now be written to yield the input impedance of the triode.

$$Z_{in} = \frac{\bar{V}_g}{\bar{I}_g} = \frac{1}{\dfrac{1}{r_g} + \dfrac{1}{1/sC_i}}.$$

This is just the parallel combination of the grid resistance r_g and a capacitor C_i. Usually $r_g \simeq \infty$ since the grid is operated with negative bias, so the input impedance is just

$$Z_{in} = \frac{1}{sC_i} = \frac{1}{s[C_{gK} + C_{gp}(1 + |A|)]}.$$

MULTISTAGE AMPLIFIERS

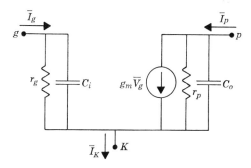

FIGURE 4-17. Equivalent circuit for a triode.

The effective grid-cathode capacitance has been increased by the term $C_{gp}(1 + |A|)$, resulting in a lower input impedance. This effect is known as the *Miller effect*.

We may use the new form of the triode equations to construct an equivalent circuit model for the triode. Application of Kirchhoff's first law to the circuit of Fig. 4-17 yields the triode equations. The term $g_m \bar{V}_g$ is called an ideal current source.

MULTISTAGE AMPLIFIERS

Let us now compute the voltage transfer function for the nth stage of amplification of a multistage amplifier (Fig. 4-18). When we have several stages

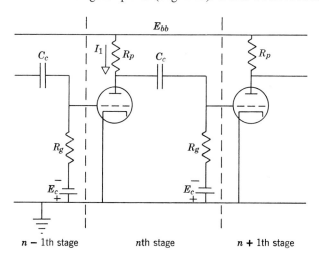

FIGURE 4-18. nth stage of a multistage triode amplifier.

127

VOLTAGE AMPLIFIERS

of amplification, we must couple one stage to another. We cannot connect the plate of one tube directly to the grid of the next since this would place the grid at high positive voltage and destroy the quiescent condition. To prevent this, we use a coupling capacitor C_c. We have also added a large grid resistor R_g to the grid circuit. The purpose of this *grid leak resistor* is to limit the flow of grid current due to the attraction of residual positive ions to the negative grid. We then have *RC* coupling, and in order not to differentiate the signals excessively, the time constant must be rather large. We then have a blocking circuit.

We may construct the equivalent circuit of Fig. 4-19. In this construction we have used the result that the supply voltage B⁺ is effectively at ground potential for fast signals. (B⁺ is at ac ground.) To see this, we write the Kirchhoff voltage law for the plate loop:

$$E_{bb} - I_1 R_p - V_{p_n} = 0.$$

The signal equation may be found quite easily by differentiation (since we are using a linear model). Hence,

$$-\Delta I_1 R_p - \Delta V_{p_n} = 0,$$

or

$$v_{p_n} = -i_1 R_p.$$

Writing the Laplace transforms, we have

$$(-\bar{I}_1) = \frac{\bar{V}_{p_n}}{R_p}.$$

The current $\bar{I}_1$ flows *to* the plate; hence, the current $(-\bar{I}_1) = \bar{V}_{p_n}/R_p$ flow *away* from the plate.

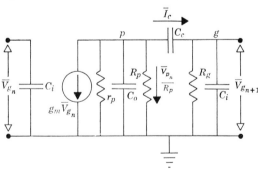

FIGURE 4-19. Equivalent circuit for the *n*th stage of a multistage triode amplifier.

MULTISTAGE AMPLIFIERS

We now solve the above circuit by writing Kirchhoff's first law at the plate node and the grid node. Hence,

$$g_m \bar{V}_{g_n} + \left(\frac{1}{r_p} + \frac{1}{R_p} + sC_0\right)\bar{V}_{p_n} + \bar{I}_c = 0;$$

$$\bar{I}_c - \left(\frac{1}{R_g} + sC_i\right)\bar{V}_{g_{n+1}} = 0.$$

The current $\bar{I}_c$ flowing through the coupling capacitor is given by

$$\bar{I}_c = sC_c(\bar{V}_{p_n} - \bar{V}_{g_{n+1}}).$$

These equations are readily solved for the voltage transfer function:

$$\bar{T}_n(s) = \frac{\bar{V}_{g_{n+1}}}{\bar{V}_{g_n}};$$

$$\bar{T}_n(s) = \frac{(-g_m r_\|)sR_g C_c}{sr_\| C_c(1 + sR_g C_i) + (1 + sr_\| C_0)[1 + sR_g(C_i + C_c)]}.$$

We may make use of the usual condition $C_c \gg C_i$ or C_o to simplify the transfer function:

$$\bar{T}_n(s) = \frac{-g_m r_\|}{1 + \dfrac{r_\|}{R_g}\left(1 + \dfrac{C_o}{C_c}\right) + sr_\|(C_i + C_o) + \dfrac{1}{sR_g C_c}},$$

or

$$\bar{T}_n(s) = \frac{-g_m}{C_\|} \cdot \frac{s}{\left[s^2 + \dfrac{1}{\tau_\|}s + \dfrac{1}{\tau_c \tau_\|}\right]},$$

where

$$\tau_c = (R_g + r_\|)C_c,$$

or

$$\tau_c \simeq R_g C_c \qquad \text{for } R_g \gg r_\|;$$

$$\tau_\| = R_\| C_\|;$$

$$C_\| = C_i + C_o;$$

$$\frac{1}{R_\|} = \frac{1}{r_\|} + \frac{1}{R_g}.$$

VOLTAGE AMPLIFIERS

There is thus a zero at the origin and two poles at the roots of the denominator. Expanding,

$$\alpha_{1,2} = -\frac{1}{2\tau_\|}\left[1 \mp \left(1 - \frac{4\tau_\|}{\tau_c}\right)^{1/2}\right];$$

$$\alpha_{1,2} = -\frac{1}{2\tau_\|}\left[1 \mp 1 \pm \frac{1}{2}\left(\frac{4\tau_\|}{\tau_c}\right) + \cdots\right];$$

$$\alpha_{1,2} \simeq \begin{cases} -\dfrac{1}{\tau_c} \\ -\dfrac{1}{\tau_\|} \end{cases},$$

since $\tau_c \gg \tau_\|$. The pole-zero plot is shown in Fig. 4-20.

We now determine the response of a typical stage to a step input:

$$\bar{V}_{g_{n+1}} = -\frac{g_m}{C_\|}\left[\frac{s}{\left(s + \dfrac{1}{\tau_\|}\right)\left(s + \dfrac{1}{\tau_c}\right)}\right]\frac{V}{s};$$

$$\bar{V}_{g_{n+1}} = AV\left[\frac{1}{s + \dfrac{1}{\tau_c}} - \frac{1}{s + \dfrac{1}{\tau_\|}}\right],$$

where

$$A = -g_m r_\|$$

and we have used the approximation $R_g \gg r_\|$. Hence,

$$V_{g_{n+1}}(t) = AV(e^{-t/\tau_c} - e^{-t/\tau_\|}),$$

or recalling that $\tau_c \gg \tau_\|$,

$$V_{g_{n+1}}(t) \simeq AV(1 - e^{-t/\tau_\|})e^{-t/\tau_c}.$$

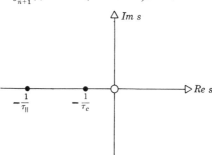

FIGURE 4-20. Pole-zero plot of the transfer function of the *n*th stage of a multistage triode amplifier.

MULTISTAGE AMPLIFIERS

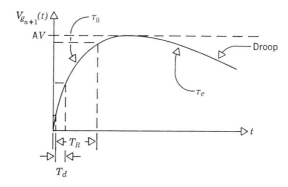

FIGURE 4-21. Output of the *n*th stage of a multistage triode amplifier for a step function input.

We see from Fig. 4-21 that the stage shapes the step input, producing a finite rise time $T_R = 2.2\tau_\|$ and that the coupling circuit causes the signal to *droop* with a time constant $\tau_c = R_g C_c$. If we had used self-biasing, the analysis would have been more complicated, but the basic result is that the cathode bypass capacitor C_K also affects the droop of pulses.

In calculating the rise time, we must also include the wiring capacitance to ground in parallel with $C_o + C_i$ giving

$$C_\| = C_i + C_o + C_\text{wiring}.$$

We note also that a delay has been introduced by the circuit. A measure of this delay is the time T_D at which the output has reached $\tfrac{1}{2}$ of its maximum value. Thus,

$$\tfrac{1}{2} = (1 - e^{-t_D/\tau_\|}),$$

or

$$T_D = 0.693\tau_\|.$$

It is instructive to examine the transfer function as a product of several factors:

$$\bar{T}_n(s) = (-g_m r_\|)\left(\frac{1}{1 + s\tau_\|}\right)\left(\frac{s\tau_c}{1 + s\tau_c}\right).$$

The first factor is just the dc gain we have calculated before. The second is the transfer function for an integrator circuit with time constant $\tau_\|$, and the last factor is the transfer function for a differentiating circuit with time constant τ_c. Looking at the pole-zero diagram, we see that if we make $\tau_c \to \infty$, the pole $-(1/\tau_c)$ moves in to the origin, where it is annihilated by the zero. The resulting transfer function describes an equivalent circuit which is just an integrating circuit with gain. This is the high-frequency equivalent circuit shown in Fig. 4-22.

VOLTAGE AMPLIFIERS

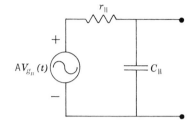

FIGURE 4-22. High-frequency equivalent circuit for an amplifier stage.

The upper half-power point for the frequency response of the amplifier stage is thus

$$f_2 = \frac{1}{2\pi r_\| C_\|}.$$

On the other hand, if we look at the limit $\tau_\| \to 0$, the pole $-(1/\tau_\|)$ moves out to $-\infty$, leaving just a differentiating circuit with gain. This is the low-frequency equivalent circuit shown in Fig. 4-23. The lower half-power point for the frequency response of the amplifier stage is thus

$$f_1 = \frac{1}{2\pi R_g C_c}.$$

Taking both limits gives us the character of the circuit in the mid-frequency range. The signal is just amplified by the mid-frequency gain $A = -g_m r_\|$. The diagram of Fig. 4-24 shows the frequency response of the amplifier stage. The *band pass* of the circuit is

$$\Delta f = f_2 - f_1.$$

If we wish to have as short a rise time T_R and delay T_D as possible, we want $\tau_\| = r_\| C_\|$ to be small. However, this is not quite the whole story since

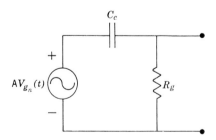

FIGURE 4-23. Low-frequency equivalent circuit for an amplifier stage.

MULTISTAGE AMPLIFIERS

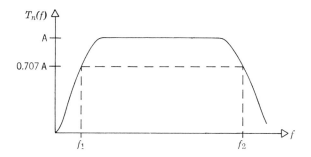

FIGURE 4-24. Frequency response of an amplifier stage.

we also want to have a reasonable gain $A = -g_m r_\|$. The gain and the upper half-power point f_2 are related; in fact, the *gain-bandwidth product* (assuming the lower half-power point f_1 to be zero) is

$$\text{GBWP} = |Af_2| = \frac{g_m}{2\pi C_\|}.$$

We begin a design by choosing a tube with as large a GBWP as possible. As a measure of the GBWP, we use the cold values of capacitance.

$$\text{GBWP} \simeq \frac{g_m}{2\pi(C_{gK} + C_{pK})}.$$

Finally, we would like to determine the output impedance of the amplifier stage. To do this we consider the high-frequency equivalent circuit of Fig. 4-25. We may replace this circuit by any circuit which gives us the same voltage at the output terminals and delivers the same current into the next

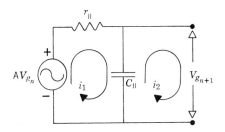

FIGURE 4-25. High-frequency equivalent circuit for an amplifier stage.

VOLTAGE AMPLIFIERS

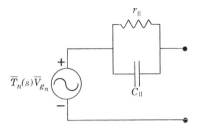

FIGURE 4-26. Another high-frequency equivalent circuit for an amplifier stage.

circuit. Writing Kirchhoff's second law for the two loops gives us

$$A\bar{V}_g - \bar{I}_1 r_{\|} - \frac{\bar{I}_1}{sC_{\|}} - \frac{\bar{I}_2}{sC_{\|}} = 0;$$

$$\bar{V}_{g_{n+1}} = \frac{\bar{I}_1 - \bar{I}_2}{sC_{\|}}.$$

We eliminate $\bar{I}_1$ from these equations and write

$$\bar{V}_{g_{n+1}} = \bar{T}_n(s)\bar{V}_{g_n} - Z_{\text{out}}(s)\bar{I}_2,$$

where

$$\bar{T}_n(s) = \frac{A}{1 + s\tau_{\|}},$$

as we expect, and

$$Z_{\text{out}}(s) = \frac{1}{\frac{1}{r_{\|}} + sC_{\|}}$$

is the output impedance. The circuit of Fig. 4-26 is thus equivalent to the one of Fig. 4-25.

IMPROVEMENT OF THE TRIODE AND FET

We have seen that the input impedance of the triode (and FET) is dominated by the capacitance between the grid (gate) and plate (drain) through the Miller effect, which multiplies the actual capacitance by the gain of the device. At high frequencies, the gain of both devices falls off badly due to this effect, and attempts were made to improve the situation.

IMPROVEMENT OF THE TRIODE AND FET

For the triode, it was obvious that if another grid were introduced between the control grid and the plate, the capacitance C_{gp} would be reduced. The new grid was called the *screen grid*, and it was operated at a positive bias so as not to disturb the existing electric field configuration in the tube. The triode thus became the *tetrode;* and, indeed, the value of C_{gp} was reduced by about an order of magnitude of so. Unfortunately, a new effect arose which vitiated the gain in time response. Even though the screen grid was very coarse and not designed to accept electrons, it was positively charged. Thus, as the plate voltage was raised, an energy would be reached at which electrons were ejected from the plate when the primary electron struck, and some of these were attracted to the screen grid, effectively *reducing* the plate current for *increasing* plate voltage (Fig. 4-27).

This region of *negative* plate resistance can be very useful, but not in a linear amplifying device. Therefore, another grid was added between the screen grid and the plate, usually with an internal connection to the *cathode*. This placed the *suppressor grid* at ground, and secondary electrons ejected from the plate would not be attracted to it or the screen grid, whose positive potential they would not feel. This cured the problem, and the resulting *pentode* became the dominant form of vacuum tube due to its capability to handle relatively fast signals.

If the screen grid were prevented from responding to the input signal by placing a large capacitor between it and ground, (the suppressor grid is already dc-coupled to ground at the cathode), our previous arguments about the functional dependence of the tube parameters still hold, since only the grid and plate vary with respect to the cathode, and all our analysis about triodes is directly applicable. Incidentally, if the grids are allowed to vary, useful effects occur, such as the mixing of two signals placed on the screen and control grids, but we will not consider such topics now.

In addition to the reduction of C_{gp} in a pentode, another effect occurs. Since the plate is now merely serving the function of attracting the electrons that make it past the other three grids, it has *little effect* on the plate current.

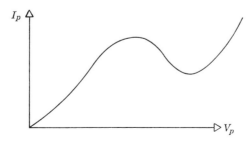

FIGURE 4-27. Plate current versus plate voltage for a tetrode.

VOLTAGE AMPLIFIERS

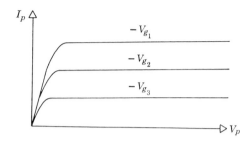

FIGURE 4-28. Plate current versus plate voltage for a pentode.

The plate current versus plate voltage curves are very flat, once all the electrons emitted at the cathode are collected by the plate (Fig. 4-28). Therefore, the plate resistance r_p becomes very large, and since the amplification μ of the tube is equal to $g_m r_p$, for an equivalent g_m the amplification will be much larger than for a triode, often 20 to 100 times greater. Therefore, the gain of a pentode is very large, resulting in high amplifications and making the application of negative feedback (described in the next chapter) highly effective in producing stability and performance impossible in low-gain tubes. It is to be noted that pentode and FET characteristics are very similar, even to the point that FETs have naturally rather low values for C_{gd} that are similar to the values of pentodes.

Despite the fact that FETs have rather good time characteristics, improvement can be effected by adding another small gate ahead of the standard two gates, resulting in a tetrode FET. Biasing into a linear region is accomplished in a standard manner through the main gates, which, being relatively large, have a much large C_{gd} than the small signal gate. High-frequency signals are applied to the signal gate, which possesses capacitive values in the order of picofarads, and results in tetrode FETs whose high-frequency input impedance matches the very large dc input impedances inherent in the device. Amplification in such devices can take place in the gigahertz region.

Experiment 4
Triode Amplifiers

The basic techniques are given here for studying tube amplifier circuits, and the procedures given now will not be repeated in such detail for later experiments.

EXPERIMENT 4

Equipment:

1. Portable volts-milliamperes-ohms meter, such as a Simpson Ohmeter Model 260
2. Pulse generator with microsecond range
3. Oscilloscope with microsecond range or better
4. Signal generator to at least 5 Mc (megahertz)

I. General

   A. Voltages up to 300 v will be used in this experiment. Before making any changes in the circuit wiring, reduce the plate voltage to zero. Use care that no bare wires or terminals are accidentally in contact before applying power to the equipment.
   B. For an accurate comparison with the theory, input and output voltages must be measured at exactly corresponding points on each stage. That is, the input voltage should be measured at the grid, not the signal generator, and the output voltage measured at the grid of the second stage.

II. Externally biased triode amplifier

   A. Quiescent conditions
   To determine quiescent conditions of the tube, that is, the conditions about which the variation produced by the input signal takes place, one uses the *average plate characteristics* chart provided by the tube manufacturer. On this chart a *load line* is drawn, using the following method. We decide upon a practical value of supply voltage E_{bb} (or B$^+$), say 300 v, and a value for the load resistor R_b (or R_L), say 10 kΩ. Then at the point where the plate current I_b is zero, the plate voltage E_p will be 300 v. This gives one intercept for the load line. At the point where E_p is zero, $I_p = E_{bb/R_b} = 300$ v$/10$ k$\Omega = 30$ ma. This gives the other intercept. A straight line is drawn between these two points; this is the load line. Now consider the coordinates of the point where the load line crosses the curve for the particular bias voltage E_c being used. These coordinates determine the quiescent conditions of the tube. For our case, $E_c = -4.5$ v, so $I_p = 11.5$ ma and $E_p = 185$ v.

   B. Selection of grid resistor R_g and coupling capacitor C_c.
   The major factor to be considered in the selection of these two components is the time constant of the two considered as a high-pass circuit. A good value for our purposes is $\tau = R_g C_c = 50$ msec. Values of R_g greater than 1 MΩ are generally undesirable, as too large a value does not allow any charge built up on the grid to leak to ground. Hence, take $R_g = 0.5$ MΩ. Then $C_c = 0.1$ μf.

VOLTAGE AMPLIFIERS

C. Circuit diagram

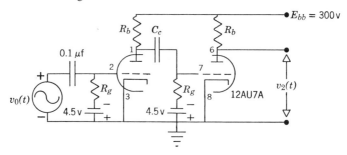

$R_g = 0.5 \text{ M}\Omega, \frac{1}{2} \text{ w}; \quad C_c = 0.1 \ \mu\text{f}, 600 \text{ v};$
$R_b = 10 \text{ k}\Omega, 2 \text{ w}; \quad V_g = -4.5 \text{ v}.$

1. Filaments: This tube can be operated on either 6.3 v or 12.6 v. For 6.3 v operation, the filaments are in parallel. To accomplish this, connect pin 4 to pin 5, and apply the 6.3 v across pin 9 and the junction of 4 and 5.
2. Construct the above circuit. All resistors and capacitors are to be wired in on the board. No decade boxes will be used. However, do not cut the leads of the resistors and capacitors to length. Batteries are to be used as the V_g.

D. Measurement of quiescent conditions

1. Measure the dc voltage from plate to cathode, using a Simpson meter (signal generator removed). Do this for each tube.
2. Measure the dc voltage across R_b, using a Simpson meter, and from this measurement compute I_p.
3. Draw the load line and find E_p and I_p graphically.
4. Compare with experimental values.
5. Determine r_p, g_m, and μ from the chart.

E. Frequency response

1. For this measurement two different signal generators will be used.
   When using the audio sig. gen., a 0.1 μf capacitor is placed in series with the output of the sig. gen. as shown, to prevent loading of the grid voltage V_g. When using the rf sig. gen., this capacitor is removed. For all measurements, the output of the sig. gen. is to be kept sufficiently low so that no distortion of the waveform is observed at the plate of the second stage. This can be checked readily with a 'scope.

EXPERIMENT 4

2. Using the ac VTVM (RMS vacuum tube voltmeter), measure the input voltage at pin 2 and the output voltage at pin 7 for several frequencies over the range 20 cycles/sec to 2 Mc/sec. ("Several" means a number sufficient to obtain a good curve when plotting.) It is best to adjust the amplitude of the sig. gen. so as to keep the input constant.
3. Plot the gain (v_2/v_0) versus frequency on 2- or 3-cycle semi-log graph paper. Determine the half-power points and the mid-frequency gain from this curve.
4. Calculate the upper and lower half-power points from the equations given in lecture and compare these with the experimental results.

   a. For this calculation it will be necessary to use the graph of tube characteristics to determine r_p, g_m, and μ.
   b. Use the following typical values for the pertinent capacitances:

   $$C_{gK} = 2.2 \text{ pf}, \quad (1 \text{ pf} = 10^{-12} \text{ f});$$
   $$C_{pK} = 1.5 \text{ pf};$$
   $$C_{gp} = 1.5 \text{ pf};$$
   $$C_{\text{wiring}} = 5.0 \text{ pf}.$$

F. Voltage

Place the audio sig. gen on the input as in part E. Set it to ~5000 c/sec. Place a Simpson dc voltmeter across the grid resistor at pin 7. Increase the input sinusoidal signal amplitude. Record the dc voltage measured by the Simpson meter. Explain why a dc voltage is developed across the grid resistor.

G. Measurement of the time constant of the circuit

1. Use the pulse generator as the source. Set the pulse width to 10 μsec, rep. rate to approximately 300 c/sec. Before connecting the pulse generator to the tube, set the pulse height as measured on the scope to 0.05 v. Connect the pulse generator to the grid of the triode, pin 2, and observe the output waveform at pin 7 with the 'scope. Measure the peak amplitude of this output pulse.
2. The time constant of the circuit is too long to allow the output pulse to charge to the full value of the output voltage, so to find the time constant τ of the amplifier, apply the formula relating τ to the charging voltage. For the full amplitude of the ouput wave, toward which the stage is "charging," use M times

VOLTAGE AMPLIFIERS

the amplitude of the input pulse, where M is the mid frequency gain of the stage.
3. From the value of τ obtained find the rise time of the stage, using $t_R = 2.2\tau$. Then, using this value of t_R and the upper half-power point f_2 obtained experimentally, verify the relation

$$t_R f_2 \approx \tfrac{1}{3}.$$

H. Output impedance
   1. Circuit diagram

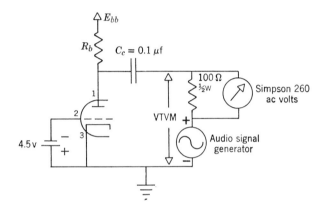

2. Set the sig. gen. to 1000 c/s and increase the output until a measurable voltage drop is obtained across the 100 Ω resistor. Read the voltage v on the VTVM and the voltage v_1 on the Simpson meter. Then the current flowing in the circuit is

$$i = \frac{v_1}{100},$$

and the magnitude of the output impedance is

$$Z = \frac{v}{i} = \frac{100}{v_1} v.$$

3. Compute the value of the output impedance as outlined in class and compare with the value obtained experimentally.

III. Self-biased triode amplifier
   A. Quiescent conditions
      1. The procedure here is basically the same as that used for the externally biased tube. One chooses a reasonable value for

EXPERIMENT 4

$R_t = R_b + R_K$ and for E_{bb}, and then one draws the load line as before, using R_t and E_{bb}. The value of V_g desired is then chosen. The intersection of the curve corresponding to this value of V_g and the load line gives I_p. From the equation

$$|V_g| = I_p R_K$$

one determines the value for R_k, and then $R_b = R_t - R_k$. For this experiment, we choose $R_t = 10$ kΩ and $E_{bb} = 300$ v. Selecting $V_g = -5$ v yields $I_p = 11$ ma. Thus, $R_K = 455$ Ω. To a sufficiently good approximation, we choose $R_K = 470$ Ω and $R_b = 10$ kΩ. To "measure" V_g, we measure the voltage drop across R_K with no signal input on the grid of the tube. Since the grid is at ground potential, the effective grid-to-cathode voltage V_g is the negative of the cathode-to-ground voltage.

2. Circuit diagram

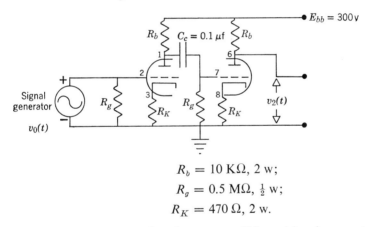

$$R_b = 10 \text{ K}\Omega, 2 \text{ w};$$
$$R_g = 0.5 \text{ M}\Omega, \tfrac{1}{2} \text{ w};$$
$$R_K = 470 \text{ }\Omega, 2 \text{ w}.$$

3. Measurement of quiescent conditions (signal generator removed). Measure the voltage drop across R_b and across R_K, using a Simpson meter. Calculate I_p and V_g from these measurements and compare with the values obtained from the load line procedure.

B. Frequency response

1. For this measurement no series capacitor is necessary with either sig. gen. Measure the input and output voltages (at pins 2 and 7, respectively), using VTVM, at several frequencies over the range from 20 c/sec to 2 Mc/sec. Follow the same general procedure as before.

VOLTAGE AMPLIFIERS

2. Plot the gain (v_2/v_0) versus frequency on 2- or 3-cycle semi-log graph paper. Determine the half-power points and the mid-frequency gain from this curve.
3. Calculate the mid-frequency gain from the equation given in class and compare with the experimental results. Also compare the half-power points with those obtained for the externally biased tube.

Experiment 5
FET Amplifiers

This experiment is the FET analog to the previous one. It includes the basic techniques for studying FET circuits and will be referenced in later FET experiments. All FET experiments will be shown with an n-channel device. To use a p-channel FET, all the polarities need to be reversed (including any diodes or electrolytic capacitors).

Equipment:
1. VTVM, or FET voltmeter with at least 5 megohm input impedance
2. Pulse generator with microsecond range
3. Oscilloscope with microsecond range or better
4. Sine-wave generator to at least 2 megahertz
5. Two (or one dual) power supplies, variable to 25 volts

I. General

    A. The FET experiments presented herein were developed using JFETs which have good resistance to overload. MOSFETs, on the other hand, are very sensitive devices with poor overload capabilities. With either variety, care must be taken to insure they are not subjected to voltages or currents above their rated values. Only under rare circumstances should the gate be forward biased. If the devices are soldered into circuits, short solder times are necessary, and a heat sink should be used on the leads as the heat is applied.

    B. Even though two FETs carry the same unit number, there can be large differences in their individual characteristics. Thus, specification sheets provided by the manufacturer can vary from a simple list of maximum and minimum values expected for a particular device to complete, multi-page dossiers containing many graphs of characteristics. We shall take the simple approach and assume that only the maximum value information is known.

EXPERIMENT 5

II. Circuit diagram

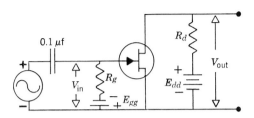

A. Construction

Construct the above circuit, using the component values discussed below unless shown above. All input measurements are to be made at the gate and output measurements made at the drain. Decade boxes may be used for R_d, or it may be wired into the circuit.

B. Component values

1. R_d is selected to limit power through the device and to insure the drain is not at ac ground. Calculate the maximum drain current from the dissipation limits for the device and choose R_d to maintain this value or less. A value of about 2000 ohms is generally a good start. Too large a value of R_d limits the current too severely.
2. R_g must be high enough to give high input impedance, but not so high as to create a bias due to the gate current. Set $R_g = 1$ megohm.

C. Measurement of quiescent conditions

1. Slowly turn up E_{dd} to about $\frac{2}{3}$ the maximum rated value for the device in use, and E_{gg} to 0.5 volt. Measure the values of V_d and V_g for no signal input. Calculate I_d. If $V_d = 0$ or E_{dd}, vary E_{gg} to obtain some voltage between these two limits.
2. Apply a small sinusoidal signal to the input. Set V_{in} to about 0.5 volt and frequency to 2 kHz.
3. Vary R_d and E_{gg} and observe V_{out} on an oscilloscope. Adjust these values to maximize V_{out} and again measure quiescent conditions after removing the input.
4. Apply the sinusoidal signal again and measure the dependence of V_{out} on V_{in}.

VOLTAGE AMPLIFIERS

D. Frequency response

1. After maximizing the gain and setting V_{in} for least distortion of V_{out} as observed on the 'scope, measure V_{out} as the frequency is changed, while maintaining V_{in} constant. Compute the gain, and plot on semi-log paper as a function of frequency. Determine the half-power points and the mid-frequency gain from the curve.
2. Use the pulse generator as the signal source and measure the time constant of the circuit. Use a pulse width of 10 μsec and repetition rate of 300 per second. Measure t_R and calculate the time constant τ. Verify, using the upper half-power point f_2 and the rise time t_R, the approximate relation $t_R f_2 \approx \frac{1}{3}$.

E. Transconductance

The FET analog to the plate resistance of a tube is the output resistance r_o. It is a good approximation that r_o is large compared to the load impedance; hence, an approximate formula for the gain of the amplifier is

$$|A| \approx g_m R_d,$$

where g_m is the transconductance. Calculate the transconductance at mid-frequency.

III. Self-biased amplifier

A. Circuit diagram

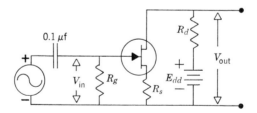

1. Construct the above circuit. Select R_s to give the same bias as used with the externally biased amplifier. Use R_d and R_g as previously determined.
2. Set E_{dd} to the same as used in the externally biased circuit and measure the quiescent conditions with no signal input. Verify that the bias is the same.

EXPERIMENT 5

3. Using the approximation

$$-|A| = \frac{g_m R_d}{1 + g_m R_s},$$

calculate the gain expected for this circuit.

4. Measure the frequency dependence of the gain of this circuit as before and compare it and the pulse response of this circuit with the previous circuit.

5. Connect a 0.1 µf capacitor in parallel with R_s and again measure the frequency response of the circuit, etc.

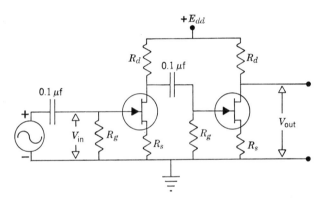

IV. Two-stage amplifier

A. Circuit diagram

B. Procedure

1. Construct the above circuit, using R_d, R_g, and R_s as in previous self-biased circuit.

2. As before, measure the gain and frequency response of the circuit.

3. Verify that the gain and upper half-power point satisfy the relations

$$A_2 = A_1^n \qquad A_1 = \text{single stage gain;}$$
$$f_{2n} = f_2 \sqrt{2^{1/n} - 1} \qquad f_2 = \text{half-power point for a single stage;}$$
$$n = \text{number of stages.}$$

145

FEEDBACK

5

INTRODUCTION

If part of the output of a system is used to modify its input, the resultant *fedback* system will possess properties strikingly different from those of the system before feedback was applied. It can be safely asserted that without feedback, much of the progress in electronics could not have occurred. Yet the concept of feedback, although recent in application, has been recognized for years in biological systems (self-regulation), mechanical systems (governors for engines), and countless others. Nevertheless, it is the extensive use of feedback in linear electronics that has allowed systems to surmount the inherent limitations present in any real device and assume performance and reliability impossible without feedback.

FEEDBACK

FIGURE 5-1. Ideal voltage amplifier.

Consider an ideal voltage amplifier with a transfer function $\bar{T}(s) = A$. Current amplifiers, although somewhat less common in practice, can be treated in a parallel manner (Fig. 5-1). Part of the output of the amplifier is now applied to the input, modifying the *voltage* that appears at the amplifier (Fig. 5-2).

Before we proceed any farther, we should be more explicit in (1) how the voltage signal applied at the amplifier is derived, and (2) how it is applied to the amplifier. There are four possibilities:

1. Voltage feedback—the reference signal is derived from the voltage $\bar{V}_2(s)$ and modified by the feedback network $\bar{\beta}(s)$.
    (a) Negative voltage feedback—the voltage-derived feedback signal *cancels* part of the input signal $\bar{V}_0(s)$.
    (b) Positive voltage feedback—the voltage-derived feedback signal *adds* to the input signal $\bar{V}_0(s)$.
2. Current feedback—the reference signal is derived from the current $\bar{I}_2(s)$ flowing at the output, and this *voltage* is then modified by the feedback network $\bar{\beta}(s)$ and applied to the amplifier.
    (a) Negative current feedback—the current-derived voltage signal *cancels* part of the input signal.
    (b) Positive current feedback—the current-derived voltage signal *adds* to the input signal.

With time-varying signals, it is possible that the system might have negative voltage feedback at one frequency and positive voltage feedback at another, since the feedback network is explicitly frequency dependent. Also, both current and voltage feedback may be simultaneously present.

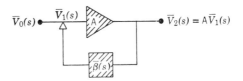

FIGURE 5-2. Ideal voltage amplifier with feedback.

NEGATIVE VOLTAGE FEEDBACK

NEGATIVE VOLTAGE FEEDBACK

Consider a circuit with voltage feedback as shown in Fig. 5-3. It is assumed that when the feedback is turned off, the voltage $\bar{V}_1(s) = \bar{V}_0(s)$, and a short circuit is placed across $\bar{V}_f(s)$. $\bar{\beta}(s)$ is the transfer function for the feedback network, so that

$$\bar{V}_f(s) = \bar{\beta}(s)\bar{V}_2(s).$$

Without feedback, we have

$$\bar{V}_2(s) = \bar{T}(s)\bar{V}_1(s)$$
$$= \bar{T}(s)\bar{V}_0(s).$$

With feedback,

$$\bar{V}_1(s) = \bar{V}_0(s) + \bar{V}_f(s)$$
$$= \bar{V}_0(s) + \bar{\beta}(s)\bar{V}_2(s).$$

Therefore, we can define a new transfer function for the system with feedback:

$$\bar{T}_f(s) = \bar{V}_2(s)/\bar{V}_0(s) = \bar{T}(s)/[1 - \bar{\beta}(s)\bar{T}(s)].$$

It is convenient to separate the frequency dependence of both the transfer function $\bar{T}(s)$ of the amplifier and the transfer function $\bar{\beta}(s)$ of the feedback network from the mid-frequency gains, so we can write

$$\bar{T}(s) = A\bar{P}(s);$$
$$\bar{\beta}(s) = \beta\bar{B}(s);$$

where $\bar{P}(s)$ and $\bar{B}(s)$ become unity in the mid-frequency range. Then

$$\bar{T}_f(s) = \left[\frac{A}{(1-\beta A)}\right]\frac{(1-\beta A)\bar{P}(s)}{[1-\beta A\bar{B}(s)\bar{P}(s)]},$$

or

$$\bar{T}_f(s) = A_f\bar{P}_f(s),$$

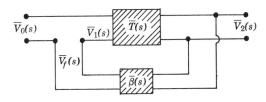

FIGURE 5-3. Voltage amplifier with feedback.

FEEDBACK

where
$$A_f = A/(1 - \beta A) = A/F$$

is the mid-frequency gain of the system with feedback, and F, defined in this equation, is called the *feedback factor*. Now we can state explicitly that positive feedback occurs when βA is greater than zero, or F less than one, and negative feedback occurs when βA is less than zero, or F greater than one.

At first glance, it may not be obvious why we wish to use negative voltage feedback, since the net amplifier gain is decreased by the introduction of the feedback loop. However, in electronics, one is always faced with *trade-offs*; and in this case, we are trading off gain for an increase in the stability of the system and/or optimization of the input and output impedance of the system.

The stability of the fedback transfer function $\bar{T}_f(s)$ against variations in the unfedback transfer function $\bar{T}(s)$ is one of the most important reasons for using negative voltage feedback. Consider the fedback transfer function

$$\bar{T}_f(s) = \frac{\bar{T}(s)}{[1 - \bar{\beta}(s)\bar{T}(s)]}.$$

Introducing a variation in $\bar{T}_f(s)$,

$$d\bar{T}_f(s) = \frac{d\bar{T}(s)}{[1 - \bar{\beta}(s)\bar{T}(s)]} + \frac{\bar{T}(s)[\bar{\beta}(s)\,d\bar{T}(s) + \bar{T}(s)\,d\bar{\beta}]}{[1 - \bar{\beta}(s)\bar{T}(s)]^2}.$$

Thus,

$$d\bar{T}_f(s)/\bar{T}_f(s) = \frac{1}{[1 - \bar{\beta}(s)\bar{T}(s)]}[d\bar{T}(s)/\bar{T}(s)] + \frac{\bar{\beta}(s)\bar{T}(s)}{[1 - \bar{\beta}(s)\bar{T}(s)]}d\bar{\beta}(s)/\bar{\beta}(s).$$

In the mid-frequency range, and assuming that $\beta A \gg 1$, this relation reduces to

$$\frac{dT_f}{T_f} = \frac{1}{F}\left(\frac{dT}{T}\right) - \frac{d\beta}{\beta}.$$

The variation in the fedback transfer function has been reduced by the feedback factor F, assuming that the feedback loop itself (which is under our control) is stable. Therefore, we have gained stability at the cost of some of the gain.

The characteristic impedances of the amplifier system will also be modified by the feedback loop. Using Thevenin's theorem, we can replace the system by a voltage source of a magnitude $\bar{T}(s)\bar{V}_1(s)$ and a series impedance $\bar{Z}(s)$ as in Fig. 5-4. Without feedback, and assuming that there is no input at the front end of the system,

$$\bar{V}(s) - \bar{I}(s)\bar{Z}(s) - \bar{T}(s)\bar{V}_1(s) = 0.$$

NEGATIVE VOLTAGE FEEDBACK

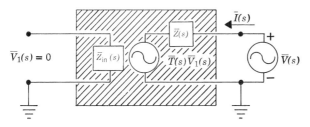

FIGURE 5-4. Voltage amplifier without feedback—Thevenin's theorem equivalent circuit.

But $\bar{V}_1(s) = 0$. Therefore,

$$\bar{V}(s)/\bar{I}(s) = Z_{out}(s) = \bar{Z}(s).$$

Now add a feedback loop to the system. $\bar{V}_1(s)$ is no longer zero even if $\bar{V}_0(s)$ is zero, under the condition that we are trying to drive current back through the system in order to determine the output impedance (Fig. 5-5). Therefore,

$$\bar{V}_1(s) = \bar{V}_f(s) = \bar{\beta}\bar{V}(s);$$
$$\bar{V}(s) - \bar{I}(s)\bar{Z}(s) - \bar{T}(s)\bar{\beta}(s)\bar{V}(s) = 0,$$

or

$$\bar{V}(s)/\bar{I}(s) = Z_{out}(s)_f = \frac{\bar{Z}(s)}{[1 - \bar{\beta}(s)\bar{T}(s)]}.$$

Again, assuming the mid-frequency range where F is constant,

$$Z_{out}(s)_f = Z(s)/F,$$

which results in an output impedance which is sharply reduced if F is large and positive, that is, negative feedback.

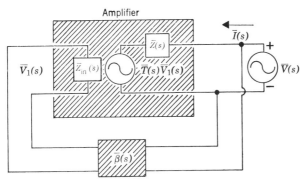

FIGURE 5-5. Voltage amplifier with feedback.

FEEDBACK

The necessity of having a very low output impedance was shown in Chapter 3, where we saw that in order to transmit a voltage pulse with minimum loss in amplitude we need a voltage source with low output impedance. Since most transmission lines have characteristic impedances in the range 50 to 125 ohms, we should have voltage sources with output impedances of this order of magnitude. We may estimate the output impedance of a typical triode amplifier stage by determining its mid-frequency value, that is, we ignore capacitive effects. Thus, $\bar{Z}_{\text{out}}(s) \simeq r_\| = r_p R_p / r_p + R_p \simeq$ several kilohms. If we wish to transmit a voltage pulse over a transmission line, we see from the above that the voltage attenuation ratio $\bar{A}(p) = Z_0 / Z_0 + \bar{Z}_1$ would be very unfavorable. To overcome this, we need an *impedance matcher*, or *buffer*, with gain unity, high input impedance, and low output impedance.

CATHODE FOLLOWER

As an example of the modification of impedance introduced by negative voltage feedback, let us analyze a circuit known as a *cathode follower* (Fig. 5-6). Note that as it is shown, the output $v_2(t)$ is applied directly to the input $v_0(t)$ through R_K. To find the open loop gain, since we cannot physically remove the feedback loop, we must remove it conceptually, simply by stating that $v_0(t) = v_g(t)$.

We will assume that i_2 and $i_g \ll i_p$, and solve this circuit by applying Kirchhoff's laws and transforming the resulting equations:

$$\bar{V}_p + \bar{I}_p R_K = 0;$$
$$\bar{V}_2 - \bar{I}_p R_K = 0;$$
$$\bar{V}_0 - \bar{V}_g = 0.$$

To this we add the triode equations in the mid-frequency approximation:

$$r_p \bar{I}_p = \mu \bar{V}_g + \bar{V}_p.$$

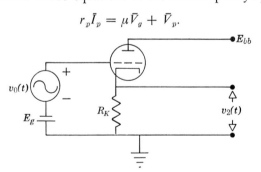

FIGURE 5-6. Cathode follower.

CATHODE FOLLOWER

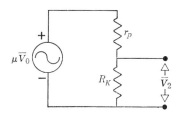

FIGURE 5-7. Equivalent circuit for a cathode follower (ignoring feedback).

Combining these equations yields

$$\bar{V}_2 = \frac{\mu R_K}{r_p + R_K} \bar{V}_0,$$

or

$$\bar{T} = \frac{\mu R_K}{r_p + R_K}.$$

Thus, the circuit acts as a voltage divider, and an equivalent circuit (if we ignore the feedback) is shown in Fig. 5-7. The output impedance is then

$$Z_{\text{out}}(s) = \frac{r_p R_K}{r_p + R_K}.$$

Now let us add feedback by applying the output at $\bar{V}_2$ directly to the input, as shown in Fig. 5-6. Therefore, instead of $\bar{V}_0 = \bar{V}_g$, we have

$$\bar{V}_0 - \bar{V}_2 = \bar{V}_g = \bar{V}_1 \text{ from Kirchhoff's laws.}$$

From the general feedback analysis,

$$\bar{V}_1 = \bar{V}_0 + \bar{V}_2$$

therefore,

$$\bar{\beta}(s) = -1,$$

and the feedback transfer function becomes

$$\bar{T}_f(s) = \frac{\bar{T}(s)}{1 - \bar{\beta}(s)\bar{T}(s)},$$

or

$$\bar{T}_f(s) = \frac{\mu R_K}{r_p + (1 + \mu)R_K} = A_f,$$

where A_f is the mid-frequency gain of the feedback system.

In the limit as μ becomes large compared to one and $\mu R_K \gg r_p$,

$$\bar{T}(s) \Rightarrow 1.$$

FEEDBACK

The fedback output impedance is found by

$$Z_{\text{out}}(s)_f = \frac{Z_{\text{out}}(s)}{F},$$

where the feedback factor is

$$F = 1 + \frac{\mu R_K}{r_p + R_K};$$

thus,

$$Z_{\text{out}}(s)_f = \frac{r_p R_K}{r_p + (1+\mu)R_K}.$$

In the same limit of large μ,

$$Z_{\text{out}}(s)_f \Rightarrow \frac{r_p}{\mu} = \frac{1}{g_m} \approx \text{few hundred ohms}.$$

We can readily add frequency dependence to this circuit through the s-dependent triode equations, and the results (see Problem 5-3) are

$$\bar{T}_f(s) = A_f \left(\frac{1 + s\dfrac{C_{gK}}{g_m}}{1 + s\tau'} \right),$$

where

$$A_f = g_m r' = \frac{\mu R_K}{r_p + (1+\mu)R_K}$$

is the mid-frequency gain, as before.

$$\tau' = r'C';$$

$$\frac{1}{r'} = \frac{1}{R_K} + \frac{1}{r_p} + g_m;$$

$$C' = C_{gK} + C_{pK}.$$

Also,

$$Z_{\text{out}}(s)_f = \frac{1}{\dfrac{1}{r'} + sC'},$$

and

$$Z_{\text{in}}(s)_f \simeq \frac{1}{sC_{gp}}$$

for a large load at $\bar{V}_2(s)$.

The pole-zero plot of the feedback transfer function in this case is shown in Fig. 5-8.

CATHODE FOLLOWER

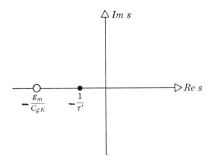

FIGURE 5-8. Pole-zero plot of the transfer function of a cathode follower.

Since $C_{pK} \simeq C_{gK}$ and $r' \simeq 1/g_m$, the zero is always about twice as far from the origin as the pole. They are not independent and, hence, cannot be made to cancel each other.

The response of the cathode follower to a step input (Fig. 5-9) illustrates how little the circuit shapes a signal:

$$V_2(t) = A_f V \left[1 - \left(1 - \frac{C_{gK}}{A_f(C_{pK} + C_{gK})} \right) e^{-t/\tau'} \right].$$

Using typical values of $r' \approx 1/g_m = 200\ \Omega$ and $C' = C_{gK} + C_{pK} \simeq 10$ pf, we find $\tau' \simeq 2$ nanosec, which is small compared with normal rise times.

As the pole is pushed out to minus infinity, the rise time becomes smaller, and the cathode follower approaches ever more closely the ideal condition for a perfect buffer in that it in no way shapes the pulse. Since C_{gp} is small, Z_{in} is very large (megohms), Z_{out} is a few hundred ohms, and the gain is approximately unity.

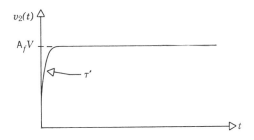

FIGURE 5-9. Output of a cathode follower for a step function input.

155

FEEDBACK

TWO-STAGE NEGATIVE VOLTAGE FEEDBACK AMPLIFIER

Let us consider two voltage amplifier stages with buffered outputs and introduce negative voltage feedback around the pair in order to achieve better stability (Fig. 5-10). For the short-time response (high frequency), we may take the single-pole approximation for the transfer function of one stage:

$$\bar{T}(s) = A\left(\frac{1}{1+s\tau}\right).$$

If the feedback loop is removed, we have, as the open-loop transfer function,

$$\bar{T}_2(s) = A^2\left(\frac{1}{1+s\tau}\right)^2.$$

Inspection of Fig. 5-10 indicates that closing the feedback loop introduces negative voltage feedback into the circuit with a feedback transfer function

$$\bar{\beta}(s) = -\frac{R}{R+R_F}.$$

(We shall ignore the capacitor C_F temporarily.)

Let us now use the general feedback relations. The open-loop amplifier gain may be written

$$\bar{T}_2(s) = A_2 \bar{P}_2(s),$$

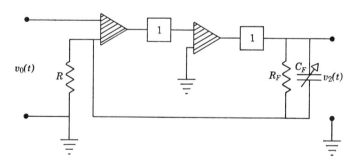

FIGURE 5-10. Two-stage voltage amplifier with negative feedback.

TWO-STAGE AMPLIFIER

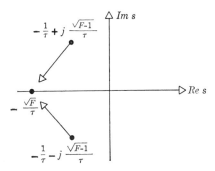

FIGURE 5-11. Pole-zero plot and trajectories of a two-stage voltage amplifier with negative voltage feedback.

and the closed-loop amplifier gain is

$$\bar{T}_f(s) = A_f \bar{P}_f(s),$$

or

$$\bar{T}_f(s) = \frac{A_2}{F}\left[\frac{F\bar{P}_2(s)}{1 - \beta A_2 \bar{B}(s)\bar{P}_2(s)}\right],$$

where

$$F = 1 - \beta A_2 = 1 + \frac{R}{R + R_F} A^2$$

is the feedback factor. We see that the mid-frequency gain of the feedback amplifier ($A_f = A_2/F$) is reduced in magnitude by the factor F, but we have traded gain for stability.

We must now investigate, however, the frequency-dependent part $\bar{P}_f(s)$ of the feedback transfer function in order to discover what effect the feedback will have upon the ability of the circuit to distort an input wave form. Noting that $\bar{B}(s) = 1$ and $\beta A_2 = 1 - F$, we have

$$\bar{P}_f(s) = \frac{F}{(1 + s\tau)^2 + (F - 1)}.$$

The poles of the feedback transfer function are seen (Fig. 5-11) to be

$$\alpha_{1,2} = \frac{1}{\tau}(-1 \pm j\sqrt{F - 1}).$$

Since we have introduced complex poles, we know that ringing will occur. Figure 5-12 illustrates the results obtained in Problem 5-5.

157

FEEDBACK

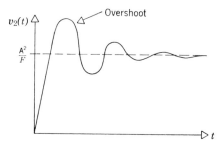

FIGURE 5-12. Output of the fedback amplifier for a step function input.

From our previous work we realize that in order to get rid of the unwanted ringing and overshoot, we must somehow move the complex poles to the real axis. Fortunately, we have some flexibility in the design of our feedback network transfer function $\bar{\beta}(s)$. If the poles are on the real axis, we must have a degenerate double-root transfer function. Examination of the feedback transfer function given above results in the following choice for $\bar{\beta}(s)$:

$$\bar{\beta}(s) = \beta(1 + as); \text{ that is,}$$

a single-zero transfer function. Substituting this into the general form of the feedback transfer function and rearranging the result gives us

$$\bar{P}_f(s) = \frac{F}{\tau^2 s^2 + [2\tau + a(F-1)]s + F}.$$

We may now make this a double-root transfer function

$$\bar{P}_f(s) = \frac{F}{(\sqrt{F} + \tau s)^2}$$

by setting
$$2\tau + a(F-1) = 2\tau\sqrt{F},$$
or
$$a = \frac{2\tau}{1 + \sqrt{F}}.$$

We have, then,
$$\bar{P}_f(s) = \frac{1}{(1 + s\tau_f)^2},$$
where
$$\tau_f = \frac{\tau}{\sqrt{F}}.$$

POSITIVE VOLTAGE FEEDBACK

The result is a doubly degenerate pole on the negative real axis and is shown in Fig. 5-11.

The question now is whether or not we can find a physical feedback network which has the transfer function given above. From Problem 5-5 we see that by simply placing a capacitor C_F across R_F and making $R \ll R_F$, that is, moving the pole of this circuit to $-\infty$, we will have the proper feedback network transfer function

$$\bar{\beta}(s) \simeq -\beta(1 + \tau_F s),$$

where

$$\tau_F = R_F C_F;$$

$$\beta = \frac{R}{R_F}.$$

The condition for compensation, that is, the elimination of the ringing, is given by

$$\tau_F = \frac{2\tau}{1 + \sqrt{F}},$$

so that the value of the compensating capacitor should be

$$C_F = \frac{1}{R_F}\left(\frac{2\tau}{1 + \sqrt{F}}\right).$$

POSITIVE VOLTAGE FEEDBACK

If a signal derived from the output voltage adds to the input voltage, the feedback is said to be positive. Consider the transfer function of a fedback system in its mid-frequency range where $\bar{T}(s) = A$, and assume for the moment that $\bar{\beta}(s) = \beta$.

$$\bar{T}_f(s) = \frac{A}{1 - \beta A}.$$

Assuming that βA is positive, as we increase from zero to a finite value, the net amplifier gain will *increase*, rather than decrease as was the case in negative voltage feedback. As βA approaches 1, the system gain becomes infinite; but long before this point, we would have violated our assumptions of linearity. However, since increases in output trigger increases in input which are amplified and returned to the output, it can be seen that the system is unstable and, in fact, may generate output independent of any input. This characteristic is utilized to design oscillators such as the Wien Bridge oscillator. All that is required is a frequency-sensitive feedback network to select

FEEDBACK

the frequency at which the system will operate and a gain such that the amplitude of the output is constant. No input is required, since random fluctuations will trigger the regenerative process and start oscillation.

It should be noted that if a system employing negative voltage feedback has a feedback loop that contains reactive elements, it is possible that at some frequency, the feedback will be shifted in phase to the point at which it becomes positive voltage feedback. Thus, oscillation may occur, which is usually undesirable.

WIEN BRIDGE OSCILLATOR

Consider the amplifier with voltage feedback shown in Fig. 5-13. Using the dc circuit scheme, we can find $\bar{\beta}(s)$:

$$\bar{\beta}(s) = \frac{\dfrac{1}{(1/R) + sC}}{\dfrac{1}{(1/R) + sC} + R + \dfrac{1}{sC}},$$

which reduces to

$$\bar{\beta}(s) = \frac{1}{\tau}\left[\frac{s}{s^2 + \dfrac{3s}{\tau} + \dfrac{1}{\tau^2}}\right]$$

where $\tau = RC$.

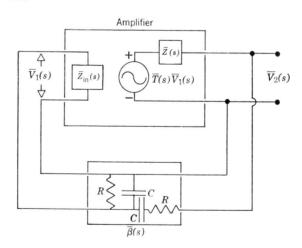

FIGURE 5-13. Wien bridge oscillator.

WIEN BRIDGE OSCILLATOR

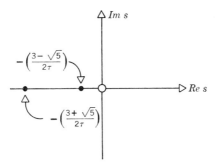

FIGURE 5-14. Pole-zero plot of the feedback network of the Wien bridge oscillator.

The pole-zero diagram for the feedback network thus has a zero at the origin and two poles on the negative real axis (Fig. 5-14), since

$$\alpha_{1,2} = \frac{-1}{2\tau}[3 \pm \sqrt{5}].$$

Using the mid-frequency gain A of the amplifier, the fedback transfer function becomes

$$\bar{T}_f(s) = \frac{A}{1 - \frac{A}{\tau}\left[\frac{s}{\left(s + \frac{3-\sqrt{5}}{2\tau}\right)\left(s + \frac{3+\sqrt{5}}{2\tau}\right)}\right]},$$

or

$$\bar{T}_f(s) = A\left[\frac{s^2 + \frac{3s}{\tau} + \frac{1}{\tau^2}}{s^2 + \frac{(3-A)s}{\tau} + \frac{1}{\tau^2}}\right].$$

This rather formidable transfer function can be best analyzed by observing the *pole-zero trajectories* as a function of the amplifier gain A (Fig. 5-15).

1. A = 0: The poles and zeros cancel, and $\bar{T}_f(s) = 0$.
2. $0 < A \leq 1$: The poles split and move toward each other on the negative real axis. The system is still stable, and the response to an input pulse would be a sum of decaying exponentials. At A = 1, the poles combine to form a double pole on the negative real axis.
3. $1 < A \leq 3$: As A continues to increase, the double pole splits and becomes complex. An input pulse would then exhibit damped ringing. As A approaches 3, the damping decreases and the frequency increases, so

FEEDBACK

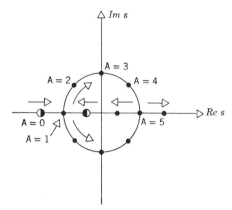

FIGURE 5-15. Pole-zero plot and trajectories of a Wien bridge oscillator.

that at A = 3, the circuit oscillates with constant amplitude at a frequency given by $1/\tau$.

4. $3 < A < \infty$: As A increases further, the frequency of oscillation starts to drop, but the damping changes to an exponential increase, which drives the amplitude out of the linear range. At A = 5, the ringing ceases as the poles converge on the positive real axis, and then split once again on the positive real axis, giving an increase that is the sum of exponentials.

For the system to operate as an oscillator, the amplifier must have a gain of 3. To maintain this condition, the amplifier will normally require negative voltage feedback.

Let us examine $\bar{\beta}(s)$ in more detail. We now know that for stable oscillation, A = 3 and $1/\tau = \omega$. At this frequency, how does the feedback network act?

$$\bar{\beta}(s) = \frac{1}{\tau}\left[\frac{s}{s^2 + \frac{3s}{\tau} + \frac{1}{\tau^2}}\right],$$

or

$$\bar{\beta}(s) = \frac{s\tau}{(s\tau)^2 + 3\tau s + 1}.$$

Let us find the "gain" of the feedback loop at resonance. For a sinusoidal signal, we can replace s by $j\omega$ in the dc circuit scheme, and obtain

$$|\beta(j\omega)| = \left|\frac{j\omega\tau}{-\tau^2\omega^2 + 3\tau j\omega + 1}\right|;$$

$$|\beta(j\omega)| = \frac{\omega\tau}{\sqrt{(1 - \omega^2\tau^2)^2 + 9(\omega\tau)^2}}.$$

PROBLEMS

At resonance, $\omega = 1/\tau$, so
$$|\beta(j\omega)| = \tfrac{1}{3}.$$

Therefore, the gain of the amplifier times the "gain" of the feedback loop equals 1, and the system operates at constant amplitude. Note also that the frequency of oscillation is more stable than the amplitude. The frequency depends only on passive elements, while the gain depends on the operation of the amplifier. In practice, negative voltage feedback is almost always used to stabilize the gain.

It is also easy to show that the phase shift introduced by going around the loop at resonance is just a multiple of 2π (see Problem 5-11).

While oscillators such as the Wien Bridge, one of a class of oscillators using passive reactive elements and feedback to achieve oscillation, are very common, better results can often be obtained by using a very stable frequency source followed by a string of amplifiers to get the desired amplitude for the output signal. The frequency stability possible in this second kind of oscillator is many orders of magnitude better than that which can be obtained even in the best passive-element amplifiers.

PROBLEMS

1. Design an amplifier with gain 100 that will suffer less than a 1% degradation in gain for a factor of 2 degradation in the operation of its active components.
2. For the amplifier considered in the previous problem, what would happen to the phase of the output signal if there were a phase shift of 30° in the unfedback amplifier? Hint: Express the signal in complex notation, and find the phase shift as a function of frequency.
3. Obtain the frequency-dependent voltage transfer function for the cathode follower. Calculate the input and output impedances.
4. Show that by the addition of the capacitor C_F, one obtains a feedback transfer function that allows for compensation in the two-stage fedback amplifier.
5. For the following circuit, show that
$$\lim_{R_2 \to 0} \bar{T}(s) \simeq \frac{R_2}{R_1}(1 + \tau_1 s),$$
where $\tau_1 = R_1 C_1$.

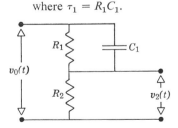

FEEDBACK

6. Show that the response of the uncompensated two-stage negative voltage feedback amplifier to a step voltage input is given by

$$v_2(t) \simeq \frac{A^2}{F}\left(1 - e^{-t/\tau_\parallel} \cos \frac{\sqrt{F}}{\tau_\parallel} t\right).$$

Note: Assume $F \gg 1$.

7. Show that the phase shift of a Wien Bridge oscillator at resonance is

$$\Delta\phi = 0.$$

8. Derive a relation that gives the value of the upper 3-dB point of an amplifier before and after the use of negative voltage feedback.

9. One of the major problems in the use of highly fedback amplifiers is that of oscillation, especially at high frequencies. Consider the following amplifier, including a stray capacitance to ground in the feedback loop. What must the value of the upper 3-dB point be so that the amplifier will not oscillate?

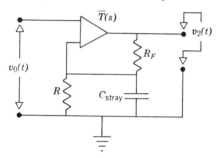

10. Show that under the conditions $R_p \gg r_p$ and $\mu \gg 1$, the gain and output impedance of the White cathode follower are given by $A \simeq 1$ and $Z_{out} \simeq 1/\mu g_m$.

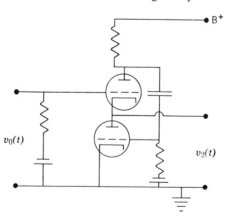

11. Find the phase shift introduced by the feedback circuit in the Wien Bridge oscillator. Compare this to the phase shift introduced in a complete loop. At

the condition of stable resonance, show that the phase shift is just

$$\Delta\phi = \pm 2n(2\pi)$$

for an n-stage amplifier in which each stage inverts the signal.

Experiment 6
The Two-Stage Pentode Negative Voltage Feedback Amplifier

Equipment:
1. Oscilloscope
2. Pulse generator with μsec range
3. Dc VTVM
4. Audio frequency generator
5. Capacitors, 10, 22, 50, and 100 pf

I. Circuit

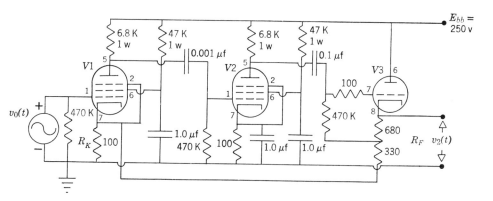

$V1 = V2 = $ 6AU6A   Filaments: 3, 4
$V3 = \frac{1}{2}$ 12AU7A   Filaments: 4–5, 9

All resistors $\frac{1}{2}$ w unless otherwise noted. Capacitors in microfarads.

II. Quiescent conditions, open loop

    A. Short pin, 1, V1, to ground.
    B. Connect R_F directly to ground, rather than to pin 7, $V1$.
    C. Using the dc VTVM, measure the voltages at pins 1, 2, 5, 6, 7 of $V1$ and $V2$, pins 6, 7, and 8 of $V3$, and the junction of the 680 Ω and 330 Ω resistors. The purpose of these measurements is to verify that the circuit is connected properly, and that the two pentodes are operating approximately as identical tubes.

FEEDBACK

   D. Using the oscilloscope, verify that the circuit is not oscillating or generating noise by placing the probe at pins 7 and 8 of $V3$.

III. Quiescent conditions, closed loop

   A. Lift R_F from ground and connect it to pin 7, $V1$.
   B. Check that the system is not oscillating by placing the scope probe at pins 7 and 8, $V3$.
   C. Disconnect R_F from pin 7, $V1$ and connect R_F to ground.

IV. Sinusoidal input, open loop

   A. Remove the ground from pin 1, $V1$, and connect the signal generator.
   B. Due to the high gain in this open-loop system, the output of the signal generator must be kept quite low to avoid overdriving the second stage.
   C. Measure the open-loop gain, defined as

   $$A_o = \frac{v_2}{v_0},$$

   at several frequencies over the range of the audio oscillator. Continue the measurement with the RF oscillator, if necessary, to reach the upper half-power point.
   D. Plot the open-loop gain versus the frequency.

V. Sinusoidal input, closed loop

   A. Lift R_F from ground and connect it to pin 7, $V1$.
   B. Measure the closed-loop gain

   $$A_c = \frac{v_2}{v_0}$$

   as above for the open-loop gain.
   C. Plot the closed-loop gain versus frequency.
   D. Using the value of open-loop gain at some mid-frequency point, calculate the expected value of closed-loop gain from

   $$A_c = \frac{A_o}{1 + |\beta| A_o},$$

   where

   $$\beta = \frac{R_K}{R_K + R_F},$$

   and compare with the experimental result.

EXPERIMENT 6

E. From the two plots of gain versus frequency, determine the change in bandwidth,

VI. Square pulse input, closed loop

A. For this part of the experiment, the pulse generator is to be used. Use a pulse rate of approximately 250 cps, and a pulse length of from 5 to 10 μsec.
B. Connect the pulse generator to pin 1 of $V1$, and observe the waveform at pin 8 of $V3$. Set the amplitude of the input pulse so that no distortion occurs at pin 8.
C. Measure the amplitude of the input pulse at pin 1, $V1$, using the oscilloscope.
D. Ringing will be observed on the output pulse at pin 8, $V3$. The following measurements are to be made:

1. Measure the period of the ringing oscillation.
2. Measure the time t from the start of the pulse to the point where the first maximum of the ringing oscillation occurs.
3. Measure the amplitude of the first maximum with respect to the baseline, and the amplitude of the main pulse (after the ringing has damped out) with respect to the baseline.

E. Using the four small capacitors provided (10, 22, 50, and 100 pf), place them one at a time across R_F while observing the waveform at pin 8, $V3$. Note qualitatively which value of capacity best removes the ringing without distorting the main pulse.
F. Calculate the total gain A, including the gain of the cathode follower, from your experimental measurement:

$$A = \frac{\text{amplitude of "main" output pulse}}{\text{amplitude of input pulse}}.$$

G. Calculate the frequency of the ringing from your experimental measurement of the period:

$$f = \frac{1}{T}.$$

H. Calculate the *fractional overshoot* obtained experimentally from fractional overshoot = $(A_{\text{1st Max}} - A_{\text{main}})/A_{\text{main}}$, where the A's denote amplitudes.

FEEDBACK

I. For the theoretical calculations, the following are the fundamental formulas:

1. $\tau_F = R_F C_F = 2\tau_{\parallel}/(1 + \sqrt{F})$
   where $\tau_{\parallel} = R_{\parallel} C_{\parallel}$;

   $R_{\parallel} \simeq R_p$

   $C_{\parallel} = C_o + C_i \simeq 20$ pf;

   $F = 1 + |\beta| A_0$, where A_0 = open-loop gain;

   $$|\beta| = \frac{R_K}{R_F + R_K}.$$

2. $v_2(t) \cong \dfrac{A^2}{F}\left[1 - e^{-t/\tau_{\parallel}} \cos\left(\dfrac{\sqrt{F}}{\tau_{\parallel}} t\right)\right].$

J. From I.2 we see that the frequency of the ringing is given by

$$f = \frac{\sqrt{F}}{2\pi\tau_{\parallel}}.$$

Calculate f and compare with the experimental value.

K. From I.1 we see that the value of capacitance C_F necessary to remove the ringing is given by

$$C_F = \frac{2\tau_{\parallel}}{R_F(1 + \sqrt{F})}$$

Caclulate C_F and compare with the value of capacity which in your opinion best removed the ringing.

L. To find the expected time to the first maximum, we differentiate I.2 and set the derivative equal to zero. From this, requiring that the time be positive, we obtain

$$t = \frac{\tau_{\parallel}}{\sqrt{F}}\left[\pi - \tan^{-1}\frac{1}{\sqrt{F}}\right].$$

Calculate t and compare with the measured value.

M. The fractional overshoot is given theoretically by

$$e^{-t/\tau_{\parallel}} \cos\left(\frac{\sqrt{F}}{\tau_{\parallel}} t\right).$$

Calculate the fractional overshoot and compare with the experimental value.

EXPERIMENT 7

Experiment 7
FET Amplifiers with Feedback

In this experiment, we use the basic two-stage FET amplifier developed previously. First, negative voltage feedback is added, then positive voltage feedback is added to produce a Wein Bridge oscillator.

Equipment:

1. Same as for the FET amplifier experiment except only one power supply (0–25 volt) needed
2. Decade resistance box

I. Negative voltage feedback

   A. Circuit diagram

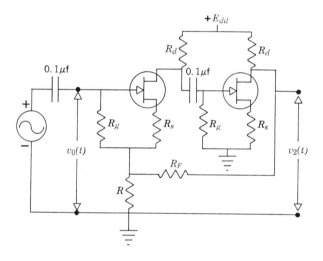

   B. Construct the above circuit with the feedback loop out of the circuit ($R = 0$, $R_F = \infty$). Use component values, as before, for the FET amplifier and verify that it still works the same. Measure the mid-frequency gain and upper half-power point.

   1. Add the feedback loop. Choose R and R_F to give a gain of 3.3 for the two-stage amplifier (to prevent loading of the output, keep $R + R_F$ to 5–10 kΩ). Use the decade box for R.
   2. Measure the gain and frequency response of the fed-back amplifier. How do these quantities depend on E_{dd}?

FEEDBACK

II. Wein Bridge oscillator

   A. Circuit diagram

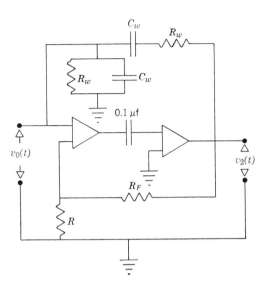

   B. Procedure

   1. As shown above, add the feedback loop consisting of R_w and C_w (two of each) to the negative fedback amplifier. The new feedback loop connects directly to the gate of the first-stage FET.
   2. Using the relation

   $$f = \frac{1}{2\pi R_w C_w},$$

   choose R_w and C_w for a convenient frequency (1–10 kHz).
   3. If the circuit does not oscillate when power is turned on, vary R until a stable output is obtained.
   4. Measure $v_0(t)$, $v_2(t)$, and the frequency of the oscillations.
   5. If R had to be varied, disconnect the oscillator feedback network (R_w and C_w) and again measure the gain of the negative fedback amplifier with this new value of R. Compute the gain expected with this value of R.

CURRENT AMPLIFIER CIRCUITS— TRANSISTORS

THINK CURRENT

If a device possesses a low input impedance, we will have to consider the current as well as the voltage of the controlling element. This prevents us from using the convenient approximation of high-input-impedance devices that the grid (gate) current is zero. It also allows us the choice of whether we wish to use current or voltage as the key variable in the analysis. Since the transistor is the dominant low-input-impedance device in use today, the choice is influenced by the basic role that current plays in the theory of the transistor, and by the small voltages that appear between certain elements in normal transistor operation (0.1 to 0.3 volt). These voltages are difficult to measure, especially without disturbing the operation of the device, while the currents normally encountered (milliamperes) times the resistances employed in transistor

CURRENT AMPLIFIER CIRCUITS

circuits (kilohms) result in voltages in volts and small probability of influencing the operation. Therefore, the theory of low-input-impedance devices will be fashioned to work for transistors (true bipolar transistors, not FETs), although it will be valid for other low-input-impedance devices.

TRANSISTORS

Basic to the bipolar transistor is the *pn* junction. We have already discussed in Chapter 4 *p*-type and *n*-type materials.* If a *p*-type semiconductor, in which holes are the majority carrier, is placed in intimate contact with an *n*-type semiconductor, in which electrons are the majority carrier, a *pn*-junction is formed. Electrons diffuse into the *p*-type region, leaving behind bound positive ions, while holes diffuse from the *p*-type region into the *n*-type region, leaving behind bound negative charges. Equilibrium is established when the potential built up across the junction by the movement of the charge is sufficient to prevent any further increase in charge transfer. The region near the junction that has reduced numbers of carriers is called the *depletion region* (Fig. 6-1).

If a potential is now applied to the junction with a polarity such that the barrier is diminished in height, the depletion layer will shrink and current flow will be established by the majority carriers. As the potential is further increased, the point is reached when the only retardation is ohmic in nature. This voltage is often only a few volts. At this point, the diode is carrying all

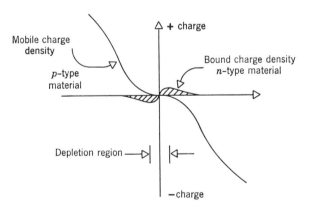

FIGURE 6-1. Bound and mobile charge near a *pn* junction.

---

* Many excellent treatments of the physics of semiconductors exist, and the Bibliography lists several which will greatly amplify this deliberately sketchy description.

TRANSISTORS

the current available from its majority carriers and is in saturation. Further increase in voltage results in no further increase in current.

If we apply a voltage with a polarity such as to enhance the barrier, charge will flow until a new equilibrium is reached, and the depletion layer will be wider than before. Few, if any, of the majority carriers will be able to surmount the barrier, and the current would go to zero except for the minority carriers. These are present in far fewer numbers than the majority carriers; but for them, the junction is forward biased and, in fact, in the saturation region. These minority carriers are highly sensitive to the temperature since they need quite a lot of thermal energy to allow them to become charge carriers. This current that appears across a back-biased *pn* junction is called the reverse saturation current, I_0.

Analysis shows that the current through a *pn* junction is related to the applied voltage by

$$I = I_0(e^{V/\eta V_T} - 1),$$

where $V_T = kT/e = 26$ mvolts;

for *Ge*, I_0 is about 10^{-6} amps $\quad \eta = 1$;

for *Si*, I_0 is about 10^{-9} amps $\quad \eta \cong 2$.

A transistor consists of two *pn* junctions in the same semiconducting bar, with the element in common being very thin. Therefore, two possible configurations exist, depending on whether the shared element (the *base*) is *n*-type or *p*-type material: *pnp* or *npn* transistors (Fig. 6-2). One of the junctions is called the *collector* and is back biased; the other is the *emitter* junction and is forward biased in normal operation. Qualitatively, the emitter injects majority carriers across the base-emitter junction into the thin, lightly-doped base region, whereupon they become minority carriers in the base. In due course, they would recombine with the majority carriers, except for the fact that the base is much thinner than the diffusion length for recombination, and most of the injected minority carriers reach the base-collector junction.

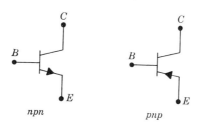

FIGURE 6-2. Schematics of *npn* and *pnp* transistors.

CURRENT AMPLIFIER CIRCUITS

This junction is back biased for the majority carriers in the base, and normally draws only a small but temperature-sensitive reverse saturation current I_0 as in a junction diode. However, it is forward biased for the now-plentiful minority carriers injected by the emitter, and they are swept across the high potential drop, gaining energy. The fraction of the emitter current that reaches the collector is called α, and this varies from about 0.90 to 0.999. The current in the collector is basically αI_E plus the reverse current mentioned above, called I_{co}:

$$I_c = \alpha I_E + I_{co}.$$

For voltage amplification, one can consider modifying the small base-to-emitter voltage, thus changing the potential barrier slightly, but producing large changes in i_e which appear as large voltage changes when this current passes through a resistor. Current amplification is accomplished because the small base current

$$I_B \simeq (1 - \alpha)I_E$$

controls the much larger emitter current. Unless base current is supplied to make up the losses of recombination, base will develop surplus charge and very quickly reach a new equilibrium potential at which only very small amounts of emitter current flow.

Finally, we have mentioned that the depletion layer varies as a function of the applied voltage. As we have seen (Fig. 2-1), there is a charge layer on either side of the junction and, consequently, an electric field across the junction. This field results in a junction capacitance, labelled C_{BC} or C_{BE}. If the junction is back biased, the depletion layer enlarges and the charge layers separate, reducing the capacity approximately as

$$C \propto 1/\sqrt{V}.$$

Therefore, $C_{BC} \ll C_{BE}$ in normal operation.

TRANSISTOR EQUATIONS

We see from the above discussion that we must take the base current into account. Thus, for the transistor we have two functional relationships or *transistor surfaces:*

$$I_C = I_C(I_B, V_C);$$
$$V_B = V_B(I_B, V_C),$$

where we have defined the variables chosen in the symbolic diagram of Fig. 6-3. Since we choose the emitter as the reference, or common, terminal and

TRANSISTOR EQUATIONS

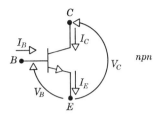

FIGURE 6-3. Voltage and current conventions in an *npn* transistor.

the base as the input terminal, we may refer to this as the common-emitter mode of operation. Of course, the base or collector could have been chosen as the common terminal.

There will now be two spaces,

$$(I_C, I_B, V_C) \quad \text{and} \quad (V_B, I_C, V_C),$$

and for a linear system we must choose a flat region on both surfaces for our operating point Q. We illustrate one of these surfaces in Fig. 6-4.

At a given temperature T_o, let us expand the above functions about a quiescent point I_{BQ}, V_{CQ}:

$$I_C = I_{CQ} + \frac{\partial I_C}{\partial I_B}\bigg|_{V_C} (I_B - I_{BQ}) + \frac{\partial I_C}{\partial V_C}\bigg|_{I_B} (V_C - V_{CQ}) + \cdots ;$$

$$V_B = V_{BQ} + \frac{\partial V_B}{\partial I_B}\bigg|_{V_C} (I_B - I_{BQ}) + \frac{\partial V_B}{\partial V_C}\bigg|_{I_B} (V_C - V_{CQ}) + \cdots ,$$

or

$$\left.\begin{array}{l} \Delta I_c \simeq \beta \Delta I_B + \dfrac{1}{r_C} \Delta V_C \\ \Delta V_B \simeq r_B \Delta I_B + h_{re} \Delta V_C \end{array}\right\} \quad \textit{transistor H-equations}$$

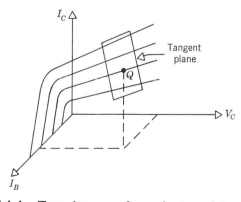

FIGURE 6-4. Transistor surface of a transistor.

CURRENT AMPLIFIER CIRCUITS

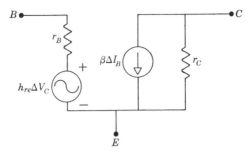

FIGURE 6-5. Equivalent circuit of a transistor.

These equations represent the tangent planes which we use to approximate the transistor surfaces for a linear analysis. As before, we use the slopes of these tangent planes as parameters representing the active device. The following are referred to as hybrid parameters since we have chosen both a current I_B and a voltage V_C as independent variables:

$$r_B = h_{ie} = \left.\frac{\partial V_B}{\partial I_B}\right|_{V_C}$$ is the common emitter input impedance, output ac short-circuited ($\Delta V_C = 0$).

$$\beta = h_{fe} = \left.\frac{\partial I_c}{\partial I_B}\right|_{V_C}$$ is the forward current transfer ratio, output ac short-circuited ($\Delta V_C = 0$).

$$\frac{1}{r_C} = h_{oe} = \left.\frac{\partial I_c}{\partial V_c}\right|_{I_B}$$ is the common emitter output admittance, input ac open-circuited ($\Delta I_B = 0$).

$$h_{re} = \left.\frac{\partial V_B}{\partial V_C}\right|_{I_B}$$ is the common emitter reverse voltage transfer ratio, input ac open-circuited ($\Delta I_B = 0$).

For typical values of these parameters, as well as formulas for converting from common-emitter to common-base or common-collector operation, see the Appendix to this chapter.

The transistor may be represented by the equivalent circuit of Fig. 6-5, since application of Kirchhoff's laws to the circuit results in the transistor hybrid equations.

BIAS CONDITIONS AND STABILIZATION

Now that we have derived the transistor hybrid equations in the linear approximation, we are in a position to design a realistic common-emitter transistor amplifier. Unfortunately, the simple biasing techniques used for

BIAS CONDITIONS AND STABILIZATION

tubes and FETs will not be adequate due to the greater variability and temperature sensitivity of transistor parameters. First, we must isolate the major sources of temperature dependence of transistors.

Recall that the total collector current I_C is the sum of a slightly attenuated emitter current plus a reverse saturation current, and this is composed of minority carriers which flow from collector to emitter, constituting a highly temperature-sensitive addition to the total current, and called $I_{CEO}(T)$ (the current from collector to emitter with the base open). We will also define a dc β, which is the ratio of the collector to the base current and an analog to the ac β which we normally use. Therefore, for an operating point Q,

$$\Delta I_C = I_C - I_{CQ}(T)$$

and

$$I_{CQ}(T_0) = \beta_{DC} I_{BQ} + I_{CEO}(T_0).$$

Now we expand $I_{CQ}(T)$ about an operating temperature T_0:

$$I_{CQ}(T) = I_{CQ}(T_0) + \left.\frac{\partial I_{CQ}}{\partial T}\right|_{T_0} (T - T_0) + \cdots.$$

The major temperature variation of I_{CQ} is due to $I_{CEO}(T)$, so that

$$I_{CQ}(T) = I_{CQ}(T_0) + \Delta I_{CEO}(T) + \cdots.$$

Therefore,

$$I_{CQ}(T) = \beta_{DC} I_{BQ} + I_{CEO}(T_0) + \Delta I_{CEO}(T).$$

Now we can redefine ΔI_C so that

$$\Delta I'_C = I_C - I_{CQ}(T_0);$$

thus, the transistor hybrid equations become

$$\Delta I'_C = \beta_{AC} I_B + \frac{1}{r_C} \Delta V_C + \Delta I_{CEO}(T)$$

and, as before,

$$\Delta V_B = r_B \Delta I_B + h_{re} \Delta V_C.$$

For the rest of this analysis, we will ignore the small term in $h_{re} \Delta V_C$.

We now have a set of hybrid transistor equations with the temperature dependence of the reverse saturation current explicitly stated. There are, of course, other sources of temperature dependence, such as $\beta_{AC}(T)$, but we will concentrate on $I_{CEO}(T)$ for the moment, since it affects the operating point Q and since we can do something about it by our choice of a biasing network.

CURRENT AMPLIFIER CIRCUITS

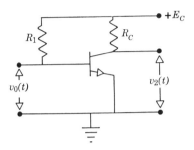

FIGURE 6-6. Simple transistor amplifier.

We might be tempted to try a simple biasing network such as the one in Fig. 6-6 to supply the base current I_B. The response of this amplifier is left as a problem for the reader but it can be stated that the parameters do not allow for much optimization, and any variation in the transistor parameters appears directly at the output.

Therefore, we will try to use feedback to reduce some of the temperature variation of the system. Since we are working on a basically current device, we will use current feedback, as shown in Fig. 6-7. Kirchhoff's equations (using total currents, dc + ac) give

$$I_B + I_C = I_E;$$

$$I_1 + I_o - I_B - I_2 = 0;$$

$$E_C = I_C R_C + V_C + I_E R_E;$$

$$E_C = I_1 R_1 + I_2 R_2;$$

$$-V_B - I_E R_E + I_2 R_2 = 0.$$

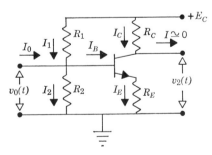

FIGURE 6-7. Stabilized transistor amplifier.

BIAS CONDITIONS AND STABILIZATION

The incremental equations now become:

$$\Delta I_B + \Delta I'_C = \Delta I_E;$$
$$\Delta I_1 + \Delta I_o - \Delta I_B - \Delta I_2 = 0;$$
$$0 = \Delta I'_C R_C + \Delta V_C + \Delta I_E R_E;$$
$$0 = \Delta I_1 R_1 + \Delta I_2 R_2;$$
$$-\Delta V_B - \Delta I_E R_E + \Delta I_2 R_2 = 0.$$

Return to the usual notation where $(I_B - I_{BQ}) = \Delta I_B = i_B$, and write in terms of i_B, i'_C, and v_C:

$$-\frac{R_2}{R_1}i_2 + i_o - i_B - i_2 = 0;$$

$$i_2 = \frac{R_1}{R_1 + R_2}(i_o - i_B);$$

$$-v_B = (i_B + i'_C)R_E - \frac{R_1 R_2}{R_1 + R_2}(i_o - i_B) = -r_B i_B.$$

Define $R_1 R_2/(R_1 + R_2) = R_{\|}$, and add the transistor hybrid equations: (In the following β will always mean β_{AC} unless otherwise noted.)

$$(R_E + R_{\|} + r_B)i_B + R_E i'_c = R_{\|} i_o;$$

$$\beta i_B - i'_C + \frac{1}{r_C}v_C = -\Delta I_{CEO}(T);$$

$$R_E i_B + (R_C + R_E)i'_C + v_C = 0.$$

Solve these three equations, and use the simplifications that $r_B = 0$ and $r_C = \infty$. Then

$$i'_C = \frac{R_{\|}}{(R_{\|} + R_E)\left[1 + \dfrac{\beta R_E}{R_E + R_{\|}}\right]}\beta i_o + \frac{\Delta I_{CEO}(T)}{1 + \dfrac{\beta R_E}{R_E + R_{\|}}}.$$

The factor $1 + [\beta R_E/(R_E + R_{\|})]$ represents the feedback from R_E of the current amplified by β, and simply is the feedback factor F:

$$i'_C = \frac{R_{\|}}{R_{\|} + R_E}\frac{\beta i_o}{F} + \frac{\Delta I_{CEO}(T)}{F}.$$

CURRENT AMPLIFIER CIRCUITS

If $R_E = 0$, then $F = 1$, and

or
$$i'_C = \beta i_o + \Delta I_{CEO}(T),$$

$$i_C = I_C - [I_{CQ}(T_0) + \Delta I_{CEO}(T)] = \beta i_o.$$

If we add the feedback factor,

$$i_C = I_C - \left[I_{CQ}(T_0) + \frac{I_{CEO}(T)}{F}\right] = \frac{R_\parallel}{R_\parallel + R_E} \frac{\beta i_o}{F}.$$

We have decreased a major source of temperature dependence, trading off gain, which is decreased by the same factor. Therefore, we have stabilized the operating point Q against variations in temperature, to guarantee that we remain in the chosen linear position of the transistor space. Variations in $\beta_Q(T)$ are not helped by this technique.

TIME RESPONSE OF TRANSISTOR CIRCUITS: COMMON-EMITTER MODE OF OPERATION

As was mentioned before, the capacitive effects in solid-state devices are more difficult to deal with than those in tubes because they are voltage dependent. However, for a linear small signal analysis, we may reasonably assume voltage-independent capacitances and adopt the following high-frequency equivalent circuit for a junction transistor. We will use the admittance model to begin with, in which the independent variables are V_B and V_C, so that we may make use of the equations previously derived for the triode. In this case, $g_m = \partial I_C/\partial V_B|_{V_C}$ (Fig. 6-8). The term involving $h_{re} = \partial V_B/\partial V_C|_{I_B}$ is small and will usually be neglected in our analysis.

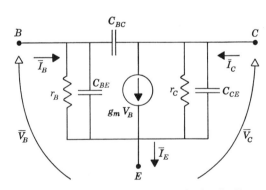

FIGURE 6-8. Transistor equivalent circuit including capacitances.

COMMON-EMITTER TIME RESPONSE

Application of Kirchhoff's nodal equations to the collector C and base B for the equivalent model shown in Fig. 6-8 results in the general active-device (in this case transistor) equations, as found for the triode in Chapter 4:

$$\bar{I}_C = \left[\frac{1}{r_C} + s(C_{CE} + C_{BC})\right]\bar{V}_C + (g_m - sC_{BC})\bar{V}_B;$$

$$\bar{I}_B = -sC_{BC}\bar{V}_C + \left[\frac{1}{r_B} + s(C_{BE} + C_{BC})\right]\bar{V}_B.$$

We could now obtain a voltage transfer function for the transistor in the same way that we did for the triode, and, indeed, transistors are sometimes used as voltage devices; however, the base current plays a more significant role in the transistor because of the transistor's inherent lower input resistance. Since the transistor is basically a current device, let us transform the transistor equations above so that the independent variables are base current $\bar{I}_B$ and collector voltage $\bar{V}_C$.

The admittance form of the transistor equations are then

$$\bar{I}_C = \bar{a}_{11}\bar{V}_C + \bar{a}_{12}\bar{V}_B;$$

$$\bar{I}_B = \bar{a}_{21}\bar{V}_C + \bar{a}_{22}\bar{V}_B.$$

We wish to transform these into the hybrid form of the transistor equations:

$$\bar{I}_C = \bar{h}_{11}\bar{I}_B + \bar{h}_{12}\bar{V}_C;$$

$$\bar{V}_B = \bar{h}_{21}\bar{I}_B + \bar{h}_{22}\bar{V}_C.$$

The relationship between the two sets of parameters may be easily shown to be

$$\bar{h}_{11} = \frac{\bar{a}_{12}}{\bar{a}_{22}};$$

$$\bar{h}_{12} = \bar{a}_{11} - \frac{\bar{a}_{12}\bar{a}_{21}}{\bar{a}_{22}};$$

$$\bar{h}_{21} = \frac{1}{\bar{a}_{22}};$$

$$\bar{h}_{22} = -\frac{\bar{a}_{21}}{\bar{a}_{22}}.$$

If we want to examine the current amplification properties of the transistor, we arrange the external circuitry so that there is a low impedance (in

CURRENT AMPLIFIER CIRCUITS

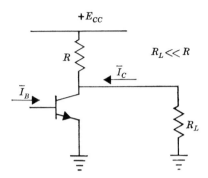

FIGURE 6-9. Transistor current amplifier.

this case, let us make it a low resistance) from collector to ac ground, as shown in Fig. 6-9 ($R_L \ll R$ or r_C).

We then have, in addition to the hybrid transistor equations, an external equation:

$$\bar{I}_C = -\frac{\bar{V}_C}{R_L}.$$

We may solve, then, for the ratio of $\bar{I}_C$ to $\bar{I}_B$ and obtain the following β-current transfer function:

$$\bar{\beta}(s) = \frac{\bar{I}_C}{\bar{I}_B} = \frac{\bar{h}_{11}}{1 + \bar{h}_{12}R_L}.$$

As the load impedance (resistance) approaches zero, we have, for the current gain from base to collector,

$$\bar{\beta}(s) \simeq \bar{h}_{11} = \frac{\bar{I}_C}{\bar{I}_B}\bigg|_{\bar{V}_C=0}.$$

From the above definitions of hybrid parameters we have

$$\bar{\beta}(s) = \frac{\beta - sr_B C_{BC}}{1 + sr_B(C_{BE} + C_{BC})},$$

where we have used the relation

$$\beta = g_m r_B.$$

The second term in the numerator will not become appreciable to β except at very high frequencies, where this model will begin to break down.

COMMON-EMITTER TIME RESPONSE

As will be seen in Chapter 7, the mode is useful for times even in the nanosecond range. Thus, we will neglect the second term in the numerator and, noting also that $C_{BE} \gg C_{BC}$, we have for the β-current transfer function of a transistor,

$$\bar{\beta}(s) = \frac{\beta}{1 + s\tau_B},$$

where

$$\tau_B = r_B C_{BE}.$$

This form applies for a transistor in the common-emitter configuration. We obtain the response of the transistor in this mode to a sinusoidal input, as before, by replacing s by $j\omega$ and rationalizing the result. Hence,

$$\hat{\beta}(\omega) = \frac{\beta}{\sqrt{1 + \omega^2 \tau_B^2}} e^{j \tan^{-1} \omega \tau_B}.$$

The upper half-power point is called the *cut-off frequency* for the transistor:

$$f_2 = f_{hfe} = \frac{1}{2\pi \tau_B}.$$

(The full name is small-signal common emitter forward current-transfer-ratio cut-off frequency.)

The *gain bandwidth product* is the frequency corresponding to $|\hat{\beta}(\omega_T)| = 1$. Thus,

$$\frac{\beta}{\sqrt{1 + \omega_T^2 \tau_B^2}} = 1;$$

$$\omega_T^2 = \frac{\beta^2}{\tau_B^2}\left(1 - \frac{1}{\beta^2}\right),$$

or

$$f_T = \frac{\beta}{2\pi \tau_B}\left(1 - \frac{1}{\beta^2}\right)^{1/2},$$

$$f_T \simeq \frac{\beta}{2\pi \tau_B};$$

$\boxed{f_T \simeq \beta f_{hfe}}$ gain × bandwidth, or GBWP (Fig. 6-10).

CURRENT AMPLIFIER CIRCUITS

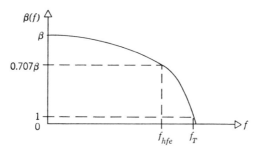

FIGURE 6-10. Frequency dependence of β.

COMMON-BASE MODE OF OPERATION

We now wish to measure voltages with respect to the base and define currents in the same way as before (Fig. 6-11). Thus,

$$\bar{V}_B = -\bar{V}_{EB};$$

$$\bar{V}_{CB} = \bar{V}_C - \bar{V}_B = \bar{V}_C + \bar{V}_{EB};$$

$$\bar{V}_C = \bar{V}_{CB} - \bar{V}_{EB}.$$

The transistor equations then become

$$\bar{I}_C = (g_m - sC_{BC})(-\bar{V}_{EB}) + \left[\frac{1}{r_C} + s(C_{BC} + C_{CE})\right](\bar{V}_{CB} - \bar{V}_{EB});$$

$$\bar{I}_E - \bar{I}_C = \left[\frac{1}{r_B} + s(C_{BC} + C_{BE})\right](-\bar{V}_{EB}) - sC_{BC}(\bar{V}_{CB} - \bar{V}_{EB}),$$

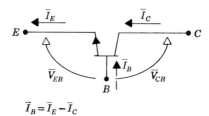

$$\bar{I}_B = \bar{I}_E - \bar{I}_C$$

FIGURE 6-11. Schematic of a transistor in common-base configuration.

COMMON-BASE MODE OF OPERATION

or

$$\bar{I}_E = -\left[g_m + \frac{1}{r_C} + \frac{1}{r_B} + s(C_{BE} + C_{CE})\right]\bar{V}_{EB} + \left[\frac{1}{r_C} + sC_{CE}\right]\bar{V}_{CB};$$

$$\bar{I}_C = -\left[g_m + \left(\frac{1}{r_C} + sC_{CE}\right)\right]\bar{V}_{EB} + \left[\frac{1}{r_C} + s(C_{BC} + C_{CE})\right]\bar{V}_{CB}.$$

These equations give us the admittance parameters for a transistor in the common-base mode of operation:

$$\bar{I}_C = \bar{g}_{11}\bar{V}_{CB} + \bar{g}_{12}\bar{V}_{EB};$$
$$\bar{I}_E = \bar{g}_{21}\bar{V}_{CB} + \bar{g}_{22}\bar{V}_{EB}.$$

Again we are interested mainly in current, so let us transform to the hybrid model for the common-base mode:

$$\bar{I}_C = \bar{b}_{11}\bar{I}_E + \bar{b}_{12}\bar{V}_{CB};$$
$$\bar{V}_{EB} = \bar{b}_{21}\bar{I}_E + \bar{b}_{22}\bar{V}_{CB}.$$

The forms of the transformation relations between coefficients are the same as before. Thus, if the load impedance is low, we have

$$\bar{\alpha}(s) = \frac{\bar{I}_C}{\bar{I}_E} \simeq \bar{b}_{11} = \frac{\bar{g}_{12}}{\bar{g}_{22}},$$

which is now the current gain from emitter to collector:

$$\bar{\alpha}(s) \simeq \frac{\bar{I}_C}{\bar{I}_E}\bigg|_{\bar{V}_{CB}=0}.$$

Thus,

$$\bar{\alpha}(s) = \frac{\left[g_m + \left(\frac{1}{r_C} + sC_{CE}\right)\right]}{\left[g_m + \frac{1}{r_C} + \frac{1}{r_B} + s(C_{BE} + C_{CE})\right]}.$$

Let us consider typical values of these parameters:

$$g_m \approx 50 \text{ millimhos};$$

$$\frac{1}{r_C} \approx \frac{1}{80 \text{ K}} \approx 0.01 \text{ millimho};$$

$$g_m \gg \frac{1}{r_C};$$

$$\frac{1}{r_B} \simeq \frac{1}{10^3} \simeq 1 \text{ millimho};$$

$$C_{BE} \simeq 100 \text{ pf}.$$

CURRENT AMPLIFIER CIRCUITS

The output capacitance is 3 to 10 pf. Hence, let us neglect $1/r_C$ and C_{CE}, giving

$$\bar{\alpha}(s) \simeq \frac{g_m}{g_m + \dfrac{1}{r_B} + sC_{BE}}.$$

Recalling that

$$\beta = r_B g_m$$

and

$$\tau_B = r_B C_{BE},$$

we have

$$\bar{\alpha}(s) = \frac{\beta}{\beta + 1 + s\tau_B};$$

$$\bar{\alpha}(s) = \frac{\beta}{1+\beta}\left[\frac{1}{\left(1 + \dfrac{s\tau_B}{1+\beta}\right)}\right];$$

$$\bar{\alpha}(s) = \frac{\alpha}{1 + s\tau_\alpha} \qquad \text{α-current transfer transfer}$$

where $\tau_\alpha = \tau_B/(1 + \beta)$, and the *emitter-collector current gain* is given by

$$\alpha = \frac{\beta}{1+\beta}.$$

The *short-circuit common-base cutoff frequency* is the upper half-power point:

$$f_\alpha = \frac{1+\beta}{2\pi\tau_B} \simeq \frac{\beta}{2\pi\tau_B} = \beta f_{hfe} = f_T;$$

$$f_\alpha \simeq \beta f_{hfe}.$$

The important point to note here is that the effective bandwidth has been increased by the factor β when the transistor is used in the common-base mode. For a single transistor, however, we do not get current gain. In Chapter 7 we discuss the *Rush transistor current amplifier*, which utilizes a second gain stage and current feedback in order to make optimum use of the available bandwidth in the input common-base transistor.

Finally, we should note that we have the following correspondence between the frequency-dependent hybrid parameters discussed above and the

APPENDIX

low- or mid-frequency parameters discussed at the beginning of this chapter:

Common-emitter mode of operation

$$\begin{pmatrix} \bar{h}_{11} & \bar{h}_{12} \\ \bar{h}_{21} & \bar{h}_{22} \end{pmatrix} \xrightarrow{\omega \to 0} \begin{pmatrix} h_{fe} & h_{oe} \\ h_{ie} & h_{re} \end{pmatrix}$$

Common-base mode of operation

$$\begin{pmatrix} \bar{b}_{11} & \bar{b}_{12} \\ \bar{b}_{21} & \bar{b}_{22} \end{pmatrix} \xrightarrow{\omega \to 0} \begin{pmatrix} h_{fb} & h_{ob} \\ h_{ib} & h_{rb} \end{pmatrix}$$

APPENDIX

Transistor hybrid parameters versus configuration for a 2N525*

Common-emitter configuration:
$h_{ie} = 1400$ ohms Input impedance 1100 ohms
$h_{re} = 3.37 \times 10^{-4}$ Output impedance 47K ohms
$h_{fe} = 44 \, (= \beta)$ Greatest power gain occurs in this configura-
$h_{oe} = 27 \times 10^{-6}$ mho tion, assuming that the load is matched to the output impedance.

Common-base configuration:
$h_{ib} = 31$ ohms Input impedance 322 ohms
$h_{rb} = 5 \times 10^{-4}$ Output impedance 322K ohms
$h_{fb} = 0.978 \, (= \alpha)$ This configuration has the lowest input
$h_{ob} = 0.6 \times 10^{-6}$ mho impedance and the highest output impedance.

Common-collector configuration:
$h_{ic} = 1400$ ohms Input impedance 48.5K ohms
$h_{rc} = 1.00$ Output impedance 1050 ohms
$h_{fc} = -45$ This configuration has the highest input
$h_{oc} = 27 \times 10^{-6}$ mho impedance and the lowest output impedance.

To be in its "active" region of operation such that a linear behavior is possible, an *npn* transistor must have its base-to-emitter voltage at about 0.5 to 0.7 volts for a silicon transistor, or about 0.1 to 0.3 volts for a germanium transistor. Reverse signs for *pnp* transistors.

* *General Electric Transistor Manual*, 7th ed., General Electric Corp., Schenectady, N.Y., p. 53.

CURRENT AMPLIFIER CIRCUITS

If one has the common-emitter hybrid parameters and wishes to convert to common-base or common-collector operation, the following approximate conversion relations will be useful:

Common-emitter to common-base

$$h_{fb} = \frac{h_{fe}}{1 + h_{fe}} \qquad h_{ob} = \frac{h_{oe}}{1 + h_{fe}}$$

$$h_{ib} = \frac{-h_{ie}}{1 + h_{fe}} \qquad h_{rb} = \frac{h_{ie}h_{oe}}{1 + h_{fe}} - h_{re}$$

Note: Some authors choose the direction for emitter current to be opposite to the choice we have made. In that case, the algebraic signs of the relations for h_{fb} and h_{ib} will be interchanged. This also leads to a negative value for α, which is equal to h_{fb}. Regardless of the choice of sign for the emitter current, the magnitudes of the common-base hybrid parameters will be given by the above relations.

Common-emitter to common-collector

$$h_{fc} = -(1 + h_{fe}) \qquad h_{oc} = h_{oe}$$

$$h_{ic} = h_{ie} \qquad h_{rc} = 1 - h_{re} \simeq 1$$

PROBLEMS

1. Using partial derivatives, show that the relation $\beta = g_m r_B$ is true.
2. Verify the transformation relations given in this chapter for the conversion between admittance and hybrid parameters.
3. Using the conditions $\beta \gg 1$, $r_C \gg r_B$, and $h_{re} \ll 1$, derive the approximate conversion relations between the common-emitter and common-base parameters given in this chapter.
4. Obtain the mid-frequency voltage gain, input impedance, and output impedance of the emitter follower (common-collector mode of operation) shown below.

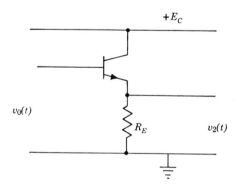

EXPERIMENT 8

5. Consider the common-collector mode of operation of a transistor. Define the appropriate hybrid parameters

$$\begin{pmatrix} h_{fc} & h_{oc} \\ h_{ic} & h_{rc} \end{pmatrix}$$

and derive the approximate conversion relations between common-emitter and common-collector hybrid parameters given in this chapter.

Experiment 8
Bias Conditions and Gain of a Transistor Amplifier

Equipment:

1. dc VTVM
2. ac VTVM
3. Oscilloscope
4. Transistor power supply
5. Audio frequency generator

I. Circuit

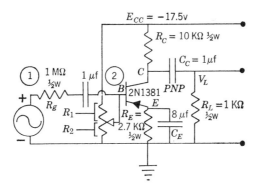

Wire the circuit as shown. The potentiometer $R_1 - R_2$ is to be set so that $I_{CQ} = 1$ ma. At this setting, $R_1 \simeq 88$ KΩ, $R_2 \simeq 12$ KΩ.

II. Dc measurements

A. With the sig. gen. off, and using a VTVM, measure the dc voltages at E_{CC}, collector, base, and emitter.
B. Verify that I_{CQ}, V_{CQ}, V_{RE}, and V_{BQ} are all consistent with the values for which the circuit was originally biased. The particular procedure used for biasing will be explained in lab.

CURRENT AMPLIFIER CIRCUITS

III. Ac measurements

  A. For the following, the ac VTVM is to be used.
  B. Set the sig. gen. to approximately 1 kc, and increase the input signal until the voltage v_1 at ① is approximately 3–5 v.
  C. Record the value of v_1 and measure and record the values of voltage v_2 at ② and v_L across R_L.
  D. Determine the value of β from the equation

  $$\beta = \frac{i_L}{i_B} \text{ (ac values)},$$

  where $i_L = v_L/R_L$ and $i_B = (v_1 - v_2)/R_g - v_2/R_{11}$; with $1/R_{11} = 1/R_1 + 1/R_2$.

  E. Vary the sig. gen. over its full frequency range, measuring the values of $v_1, v_2,$ and v_L at several frequencies. Plot β as a function of frequency, showing the half-power points. Also determine the GBWP, using the fact that the GBWP = the frequency at which $\beta = 1$.

IV. Transistor Curve-Tracer

A transistor curve-tracer, if available, is extremely useful for this experiment. A talk on its operation is usually given at this time.

AMPLIFIERS WITH SEVERAL ACTIVE ELEMENTS

INTRODUCTION

We are now in a position to analyze more complex systems involving two or more active elements. It is obvious that the choice of circuits to analyze is extremely large, so we shall choose circuits that each illustrate a particular aspect of multi-element devices. The first circuit, a Darlington pair, will be treated as a single transistor with very useful parameters. The second, a difference amplifier, is a case in which the two elements join to give an output far different from each one's output by itself. Finally, we shall treat circuits, such as the Rush current amplifier, that illustrate ways of obtaining maximum bandwidth from a transistor, and also mention changes that occur when systems are cascaded to give gain impossible with a single stage. For simplicity, we will use midfrequency analysis in the first two examples.

AMPLIFIERS WITH ACTIVE ELEMENTS

THE DARLINGTON CIRCUIT

We will approach the transistor pair (Fig. 7-1) as a single transistor, whose parameters will be written as primes, such as β'. First, we shall calculate the new hybrid parameters in an ac analysis, under the assumption that $r_C = \infty$, and $h_{re} = 0$. Thus, for each transistor,

$$\beta = \frac{i_C}{i_B};$$

$$\alpha = \frac{i_C}{i_E} = \frac{\beta}{1 + \beta};$$

$$r_B = \frac{v_B}{i_B}.$$

We have for the separate transistor, then,

$$\beta_1 = \frac{i_{C1}}{i_{B1}}; \qquad \frac{i_{C1}}{i_{E1}} = \frac{\beta_1}{1 + \beta_1};$$

$$\beta_2 = \frac{i_{C2}}{i_{B2}}; \qquad \frac{i_{C2}}{i_{E2}} = \frac{\beta_2}{1 + \beta_2}.$$

Therefore,

$$\frac{i_{B1}}{i_{E1}} = \frac{1}{1 + \beta_1} \quad \text{and} \quad \frac{i_{B2}}{i_{E2}} = \frac{1}{1 + \beta_2},$$

or, since

$$i_{B2} = i_{E1}, \qquad \frac{i_{B2}}{i_{E2}} = \frac{i_{E1}}{i_{E2}}.$$

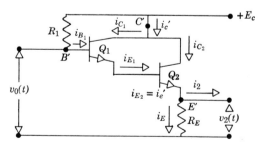

FIGURE 7-1. Darlington circuit.

THE DARLINGTON CIRCUIT

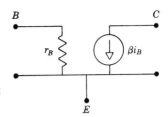

FIGURE 7-2. Simple transistor equivalent circuit.

Now

$$\beta' = \frac{i'_C}{i'_B} = \frac{i_{C1} + i_{C2}}{i_{B1}}$$

$$= \frac{\beta_1 i_{B1} + \beta_2 i_{E1}}{i_{B1}},$$

or

$$\beta' = \beta_1 + \beta_2 + \beta_1\beta_2.$$

Now let us calculate the equivalent base resistance r'_B (Fig. 7-2):

$$r'_B = \frac{v'_B}{i'_B} = \frac{v_{B1} + v_{B2}}{i_{B1}}.$$

But

$$v_{B1} = r_{B1} i_{B1};$$

$$v_{B2} = r_{B2} i_{B2} = r_{B2} i_{E1}.$$

Thus,

$$v_{B2} = r_{B2}(1 + \beta_1) i_{B1},$$

or

$$r'_B = r_{B1} + (1 + \beta_1) r_{B2}.$$

We can now analyze the much simpler circuit of Fig. 7-3. First, R_1 is a large resistor used for biasing purposes, and we will ignore it for the time

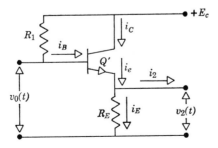

FIGURE 7-3. Equivalent circuit for a Darlington circuit.

AMPLIFIERS WITH ACTIVE ELEMENTS

being. Applying Kirchhoff's laws,

$$i_B + i_C = i_E + i_2 = i_e;$$
$$v_0 = v_B + i_E R_E;$$
$$v_C = -i_E R_E;$$
$$v_2 = i_E R_E.$$

Add the transistor hybrid equation in simplified form:

$$i_C = \beta' i_B;$$
$$v_B = r'_B i_B;$$
$$\frac{i_C}{i_e} = \frac{\beta'}{1 + \beta'}.$$

Therefore, the current gain of the system, defined as the current i_2 through a short-circuit load divided by the input current, is

$$A_I = \frac{i_e}{i_B} = 1 + \beta'.$$

The voltage gain is the voltage v_2 with an infinite load, or

$$A_V = \frac{v_2}{v_0}\bigg|_{i_2=0} = \frac{i_E R_E}{v_B + i_E R_E}$$
$$= \frac{(\beta' + 1) i_B R_E}{r'_B i_B + i_B(\beta' + 1) R_E}$$
$$= \frac{(1 + \beta') R_E}{r'_B + (1 + \beta') R_E}$$
$$\approx 1 \qquad \text{for } (1 + \beta') R_E \gg r'_B.$$

To find the output impedance, we can use the short-circuit theorem, since we already have i_{short} and v_{open}:

$$Z_{\text{out}} = \frac{v_{\text{open}}}{i_{\text{short}}}$$
$$= \frac{\left(\dfrac{v_0(1 + \beta') R_E}{r'_B + (1 + \beta') R_E}\right)}{i_B(1 + \beta')}.$$

194

THE DARLINGTON CIRCUIT

Since $v_0 \approx v_B$ for the short circuit case, we can use $v_B/i_B = r'_B$. Therefore,

$$Z_{\text{out}} = \frac{r'_B R_E}{r'_B + (1 + \beta')R_E},$$

or, since $(1 + \beta')R_E \gg r'_B$,

$$Z_{\text{out}} \simeq r'_B/\beta' \simeq r_{B2}/\beta_2.$$

For example, using the values for a 2N525 transistor, $\beta \simeq 44$, $r_B \simeq 1400\ \Omega$:

$$Z_{\text{out}} \simeq \frac{1400\ \Omega}{44} \simeq 32\ \Omega.$$

For the input impedance,

$$Z_{\text{in}} = v_0/i_B.$$

Assume the load is much larger than R_E, so that $i_2 \simeq 0$. Then $i_e = i_E$ and

$$Z_{\text{in}} = \frac{v_0}{i_B} = \frac{v_B + i_E R_E}{i_B}$$

$$= \frac{r'_B i_B + (1 + \beta')i_B R_E}{i_B}$$

$$= r'_B + (1 + \beta')R_E,$$

or

$$Z_{\text{in}} \simeq \beta_1 r_{B2} + \beta_1 \beta_2 R_E.$$

If $R_E \simeq 10\ \text{k}\Omega$, $r_{B2} \simeq 1.4\ \text{k}\Omega$, and $\beta \simeq 44$,

$$Z_{\text{in}} \simeq \beta^2 R_E \simeq 20 \times 10^6\ \Omega.$$

This analysis, performed with greater care (since on this scale $r_C \neq \infty$), would result in a value closer to a few $\times 10^6\ \Omega$'s. In addition, we have ignored R_1, which would normally be $\sim 10^5\ \Omega$'s, and would be in parallel with the Darlington Z_{in}. However, the point is clear that the Darlington configuration is an excellent buffer stage, since

$$A_V \simeq 1;$$

$$A_I \simeq \beta^2;$$

$$Z_{\text{in}} \simeq 10^6\ \Omega;$$

$$Z_{\text{out}} \simeq r_B/\beta \simeq 30\ \Omega.$$

These values should be compared with the cathode follower, Chapter 5.

The Darlington connection allows us to obtain the maximum input impedance possible from a transistor. Such Darlington pairs are available

AMPLIFIERS WITH ACTIVE ELEMENTS

encapsulated as a single transistor and, of course, in the form of an integrated circuit, where the whole circuit is formed on a single chip of silicon or germanium.

The reader should now reconsider the above analysis of the input and output impedances for a simple emitter follower and a Darlington circuit for the case of finite source and load impedances.

A DIFFERENCE AMPLIFIER

A second example of the use of two active elements is the difference amplifier. In this case, we will obtain an amplifier that amplifies the *difference* of two input signals. Minor modifications of the circuit will also result in other functions performed by the same basic circuit. Each different function, unfortunately, leads to a different name for the same basic circuit.

We will solve the circuit of Fig. 7-4, using two independent inputs $v_{01}(t)$ and $v_{02}(t)$. Apply Kirchhoff's laws to obtain the signal equations, using the following assumptions:

$$R_B = \infty;$$
$$i_E = \alpha i_C \simeq i_C;$$
$$i_2 = 0;$$
$$\beta_1 = \beta_2, \quad r_{B1} = r_{B2}, \quad r_{C1} = r_{C2};$$
$$h_{re1,2} = 0.$$

Then
$$v_{21} = -i_1 R_1;$$
$$v_{22} = -i_2 R_2;$$
$$i_1 R_1 + v_{C1} + (i_1 + i_2) R_E = 0;$$
$$i_2 R_2 + v_{C2} + (i_1 + i_2) R_E = 0;$$
$$v_{01} - v_{B1} - (i_1 + i_2) R_E = 0;$$
$$v_{02} - v_{B2} - (i_1 + i_2) R_E = 0.$$

Add to these the transistor hybrid equations in the mid-frequency range, assuming that the transistors are identical:

$$i_C = \beta i_B + \frac{1}{r_C} v_C;$$
$$v_B = r_B i_B.$$

A DIFFERENCE AMPLIFIER

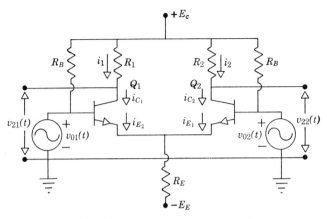

FIGURE 7-4. Difference amplifier.

We can solve these equations by finding i_1 and i_2 through any of a number of methods. When there are this many variables, use of matrix methods can help keep track of the parameters. We will use the determinant simply to aid in cancelling variables. Cramer's rule, given in most references on mathematical methods, is probably the easiest to use, but we will illustrate a simpler method here.

v_{B1}	v_{B2}	v_{C1}	v_{C2}	i_1	i_2	
1	0	0	0	R_E	R_E	v_{01}
0	1	0	0	R_E	R_E	v_{02}
$\beta \dfrac{r_C}{r_B}$	0	1	0	$-r_C$	0	0
0	$\beta \dfrac{r_C}{r_B}$	0	1	0	$-r_C$	0
0	0	1	0	$(R_1 + R_E)$	R_E	0
0	0	0	1	R_E	$(R_2 + R_E)$	0

The symmetry of the amplifier makes this technique as easy as any in this case. Adding the rows, multiplied by factors chosen to result in the elimination of v_{B1}, v_{B2}, v_{C1}, and v_{C2}, gives

$$v_{22}(t) = i_2 R_2 = \frac{\dfrac{r_C}{r_B}\beta\left(1 + \dfrac{r_C}{r_B}\beta\right) R_2 R_E(v_{02}(t) - v_{01}(t)) + \dfrac{r_C \beta}{r_B} R_2(R_1 + r_C)v_{02}(t)}{(R_1 + r_C)(R_2 + r_C) + \left(1 + \dfrac{r_C \beta}{r_B}\right) R_E[(R_1 + r_C) + (R_2 + r_C)]}.$$

For $v_{21}(t)$, interchange all the 1's and 2's.

AMPLIFIERS WITH ACTIVE ELEMENTS

At this point, we can concentrate on this basic configuration. First, the difference amplifier, for which $R_1 = R_2 = R$, and the output is taken at $v_{22}(t)$ or $v_{21}(t)$:

$$v_{22}(t) = \frac{\left(\frac{\beta r_C}{r_B}\right) R}{2(R + r_C)} \left[\frac{v_{02}(t) - v_{01}(t) + 2av_{02}(t)}{1 + a}\right],$$

where

$$a = \frac{R + r_C}{2\left(1 + \frac{\beta r_C}{r_B}\right) R_E}.$$

Add and subtract $av_{01}(t)$ from the numerator:

$$v_{22}(t) = \frac{\beta R r_C}{2r_B(R + r_C)} \left\{[v_{02}(t) - v_{01}(t)] + \frac{a}{(1 + a)}[v_{02}(t) + v_{01}(t)]\right\}.$$

We have now reduced the output to the form of the difference of the two inputs and the sum of the two inputs. To obtain a true difference amplifier that responds to the difference of the two signals, we want to eliminate the sum signal, which means to minimize $a/(1 + a)$. Writing

$$v_{02}(t) - v_{01}(t) = v_d(t) \qquad \text{the difference mode;}$$

$$\frac{v_{02}(t) + v_{01}(t)}{2} = v_s(t) \qquad \text{the common, or sum, mode,}$$

then

$$v_{\text{out}}(t) = \frac{\beta R r_C}{2r_B(R + r_C)} \left[v_d(t) + \frac{2a}{1 + a} v_s(t)\right].$$

The factor $(1 + a)/2a$ is called the *common-mode rejection ratio*, or CMRR.

In most cases, $r_C \gg R$ and $\beta \gg 1$, so that the result becomes

$$v_{\text{out}}(t) \simeq \frac{\beta R}{2r_B} \left[v_d(t) + \frac{v_s(t)}{\text{CMRR}}\right],$$

where the CMRR

$$\simeq \frac{1}{2a} \simeq \frac{\beta R_E}{r_B} \gg 1.$$

For reasonable values, $\beta = 44$, $R_E = 10$ K, $r_B = 1.4$ K,

$$\text{CMRR} \simeq 340,$$

or the sum signal is present only to about 0.3%. By raising R_E and choosing

RUSH TRANSISTOR CURRENT AMPLIFIER

a higher β, this can be considerably improved, and, in fact, CMRR's on the order of 10^6 can be attained. It should be noted that if the two transistors are in a similar temperature environment, such as in a single integrated circuit chip, that deviations from the operating point Q will represent a common-mode signal and be suppressed. This convenient fact is used to make an amplifier stabilized against changes in the operating point Q, simply by grounding one input.

RUSH TRANSISTOR CURRENT AMPLIFIER*

In Chapter 6 we pointed out that in order to obtain the full bandwidth and, hence, shortest rise time, a transistor should be used in its common-base configuration. We would now like to discuss a circuit which utilizes this configuration and current feedback to produce a very fast current amplifier, that is, in the nanosecond range. We note also that, by using current amplification, we are able to overcome the shunting capacitive effects in voltage amplifiers which limit the upper half-power point or frequency response of the amplifier. The amplifier stage is shown in Fig. 7-5 and consists of a complementary pair, that is one npn transistor and one pnp transistor, with common-base input and negative current feedback.

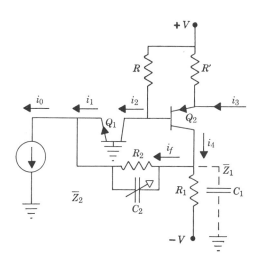

FIGURE 7-5. Rush transistor current amplifier.

* Charles J. Rush, "New Technique for Designing Fast Rise Transistor Pulse Amplifiers," *Review of Scientific Instruments* **35**, 149 (1964). This reference uses Laplace transforms.

AMPLIFIERS WITH ACTIVE ELEMENTS

Since we are dealing with current, we set up a nodal analysis. First, let us assume that R and R' are large compared with other impedances in the circuit, so that we may neglect the currents flowing through them. We also assume an ideal current source and zero impedance load. Our goal is to calculate the current transfer function:

$$\bar{\beta}(s) = \frac{\bar{I}_3(s)}{\bar{I}_0(s)}.$$

To solve for $\bar{\beta}$, we write the transistor equations and the current nodal equations:

$$\left.\begin{aligned} \bar{I}_2 &= \bar{\alpha}_1 \bar{I}_1 \\ \bar{I}_4 &= \bar{\alpha}_2 \bar{I}_3 \\ \bar{I}_3 &= (\bar{\beta}_2 + 1)\bar{I}_2 \end{aligned}\right\} \quad \text{transistor equations}$$

$$\bar{I}_1 + \bar{I}_f = \bar{I}_0 \quad \text{nodal equation}$$

$$\bar{I}_f = \frac{Z_1}{Z_1 + Z_2} \bar{I}_4 \quad \text{current divider equation}$$

These equations may then be immediately solved for the ratio $\bar{I}_3/\bar{I}_0$, giving

$$\bar{\beta}(s) = \frac{\bar{\alpha}_1(\bar{\beta}_2 + 1)}{1 + \bar{\alpha}_1 \bar{\beta}_2 \dfrac{Z_1}{Z_1 + Z_2}},$$

where we have made use of the relation

$$\bar{\alpha}_2 = \frac{\bar{\beta}_2}{\bar{\beta}_2 + 1}.$$

We note that in the numerator we have the forward current gain $\bar{\alpha}_1(\bar{\beta}_2 + 1)$, and in the denominator the loop gain $\bar{\alpha}_1 \bar{\beta}_2 Z_1/(Z_1 + Z_2)$.

The current transfer function for the common-base stage Q_1, derived in Chapter 6, is

$$\bar{\alpha}_1 = \frac{\alpha_1}{1 + s\tau_\alpha},$$

where

$$\alpha_1 = \frac{\beta_1}{\beta_1 + 1};$$

$$\tau_{\alpha_1} = \frac{\tau_{B1}}{1 + \beta_1};$$

$$\tau_{B1} = r_{B1} C_{BE1}.$$

RUSH TRANSISTOR CURRENT AMPLIFIER

Since Q_2 is operated in the common-emitter mode for the feedback current $\bar{I}_f$, we have

$$\tilde{\beta}_2 = \frac{\beta_2}{1 + s\tau_{B2}},$$

where

$$\tau_{B2} = r_{B2} C_{BE2}.$$

The current-divider ratio is

$$\frac{Z_1}{Z_1 + Z_2} = \frac{\dfrac{1}{(1/R_1) + sC_1}}{\dfrac{1}{(1/R_1) + sC_1} + \dfrac{1}{(1/R_2) + sC_2}};$$

$$\frac{Z_1}{Z_1 + Z_2} = \frac{R_1}{R_1 + R_2}\left(\frac{1 + s\tau_2}{1 + s\tau_\|}\right),$$

where

$$\tau_2 = R_2 C_2;$$
$$\tau_\| = R_\| C_\|;$$
$$\frac{1}{R_\|} = \frac{1}{R_1} + \frac{1}{R_2};$$
$$C_\| = C_1 + C_2.$$

We may simplify the circuit by using pole-zero cancellation on the current divider to obtain the condition of compensation:

$$\tau_2 = \tau_\|, \quad \text{or} \quad R_1 C_1 = R_2 C_2.$$

We then have a frequency-independent feedback network, since

$$\frac{Z_1}{Z_1 + Z_2} = \frac{R_1}{R_1 + R_2}.$$

With the above substitutions, the current transfer function becomes

$$\tilde{\beta}(s) = \frac{\dfrac{\alpha_1}{1 + s\tau_{\alpha_1}}\left(\dfrac{\beta_2}{1 + s\tau_{B2}} + 1\right)}{1 + \dfrac{R_1}{R_1 + R_2}\left(\dfrac{\alpha_1}{1 + s\tau_{\alpha_1}}\right)\left(\dfrac{\beta_2}{1 + s\tau_{B2}}\right)},$$

or

$$\tilde{\beta}(s) = \frac{\alpha_1}{\tau_{\alpha_1}}\left[\frac{s + \dfrac{\beta_2 + 1}{\tau_{B2}}}{s^2 + \left(\dfrac{\tau_{\alpha_1} + \tau_{B2}}{\tau_{\alpha_1}\tau_{B2}}\right)s + \dfrac{1}{\tau_{\alpha_1}\tau_{B2}}\left(1 + \dfrac{\alpha_1 \beta_2 R_1}{R_1 + R_2}\right)}\right].$$

AMPLIFIERS WITH ACTIVE ELEMENTS

Thus, we have a zero at $-(\beta_2 + 1)/\tau_{B_2}$ and poles at

$$s_{1,2} = -\frac{\tau_{\alpha_1} + \tau_{B_2}}{2\tau_{\alpha_1}\tau_{B_2}} \left\{ 1 \pm \left[1 - \frac{4\tau_{\alpha_1}\tau_{B_2}}{(\tau_{\alpha_1} + \tau_{B_2})^2} \frac{(1 + \alpha_1\beta_2 R_1)}{R_1 + R_2} \right]^{1/2} \right\}.$$

By adjusting the feedback ratio $R_1/(R_1 + R_2)$, one can have the overdamped, critically damped, or underdamped case; that is, a conjugate pair of complex poles, a double real pole, or two real poles. For each value of the ratio $R_1/(R_1 + R_2)$, one must readjust the current-divider compensation condition. Here again we choose the critically damped case by moving the poles to the negative real axis to give us the following double-pole transfer function:

$$\bar{\beta}(s) = \frac{\alpha_1}{\tau_{\alpha_1}} \left[\frac{s + \dfrac{\beta_2 + 1}{\tau_{B_2}}}{\left(s + \dfrac{\tau_{\alpha_1} + \tau_{B_2}}{2\tau_{\alpha_1}\tau_{B_2}}\right)^2} \right].$$

The condition that must be satisfied for the critically damped case is

$$1 - \frac{4\tau_{\alpha_1}\tau_{B_2}}{(\tau_{\alpha_1} + \tau_{B_2})^2} \left[\frac{(1 + \alpha_1\beta_2 R_1)}{R_1 + R_2} \right] = 0.$$

We may simplify these results by noting that $\alpha_1 \simeq 1$, $\beta_1 \gg 1$, and that $\tau_{\alpha_1} = \tau_{B_1}/(1 + \beta_1) \ll \tau_{B_2}$ since $\tau_{B_1} \simeq \tau_{B_2}$. Making use of these approximations and writing the results in terms of the *gain-bandwidth products* f_T of the transistors gives us

$$\bar{\beta}(s) = \frac{4f_{T_2}}{f_{T_1}} \left[\frac{1 + \dfrac{s}{2\pi f_{T_2}}}{\left(1 + \dfrac{s}{\pi f_{T_1}}\right)^2} \right],$$

where

$$f_{T_1} = \frac{\beta_1}{2\pi\tau_{B_1}};$$

$$f_{T_2} = \frac{\beta_2}{2\pi\tau_{B_2}}.$$

To further simplify our analysis and to make it possible to cascade stages, let us put a common-base current buffer after the fedback pair. If we

RUSH TRANSISTOR CURRENT AMPLIFIER

choose $f_{T_3} \simeq f_{T_2}$, then the common-base single-pole transfer function

$$\bar{\beta}_3(s) = \frac{\alpha_3}{1 + \dfrac{s}{2\pi f_{T_3}}} \simeq \frac{1}{1 + \dfrac{s}{2\pi f_{T_3}}}$$

may be used to cancel the zero of the fedback pair. Thus, the total current transfer function is

$$\bar{\beta}_T(s) = \frac{4 f_{T_2}}{f_{T_1}} \left[\frac{1}{\left(1 + \dfrac{s}{\pi f_{T_1}}\right)^2} \right].$$

To find the rise time, we may consider this transfer function to be the result of the product of two identical stages,

$$\bar{\beta}_1(s) = 2\sqrt{\frac{f_{T_2}}{f_{T_1}}} \left(\frac{1}{1 + \dfrac{s}{\pi f_{T_1}}} \right),$$

the rise time of which is

$$T_{R_1} = 2.2 \left(\frac{1}{\pi f_{T_1}} \right).$$

For two cascaded identical stages (see next section), we would then have

$$T_R = \frac{\sqrt{2}\,(2.2)}{\pi f_{T_1}}.$$

Thus, the rise time and gain of the Rush transistor current amplifier are given by

$$T_R = \frac{1}{f_{T_1}};$$

$$A = 4 \frac{f_{T_2}}{f_{T_1}}.$$

The condition for critical damping, for the case of high loop gain, gives us the additional design equation

$$\frac{R_1 + R_2}{R_1} \simeq A.$$

As indicated in the discussion of cascaded stages, the overall gain A_T and overall rise time T_{RT} of several stages of amplification may be found by

AMPLIFIERS WITH ACTIVE ELEMENTS

multiplying the gain of the individual stages and taking the square root of the sum of squares of the rise times of individual stages.

Using the above analysis, Rush has designed and built a fast transistor amplifier having a gain of 1580 and rise time of 3 nsec.

CASCADED AMPLIFIERS

Before leaving this chapter, we would like to discuss some general results which are applicable to cascaded amplifier stages. We consider the general circuit of Fig. 7-6.

If each identical amplifier stage is represented by a single-pole transfer function (corresponding to the short-time or high-frequency response)

$$\bar{T}(s) = \frac{A}{1 + s\tau},$$

then the overall transfer function will be just the product transfer function:

$$\bar{T}_N(s) = \frac{A^N}{(1 + s\tau)^N}.$$

To see the effect of cascading stages, let us assume voltage amplifiers and determine the response of the system to a step voltage input. The output voltage transfer function would be

$$\bar{V}_N(s) = A^N V \left[\frac{1}{(1 + s\tau)^N} \left(\frac{1}{s} \right) \right].$$

Looking up the inverse transform, we have

$$v_N(t) = A^N V \left\{ 1 - \left[1 + t/\tau + \frac{(t/\tau)^2}{2!} + \frac{(t/\tau)^3}{3!} + \cdots + \frac{(t/\tau)^{N-1}}{(N-1)!} \right] e^{-t/\tau} \right\},$$

which is shown in Fig. 7-7.

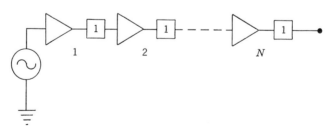

FIGURE 7-6. Cascaded amplifiers.

GAUSSIAN AMPLIFIER

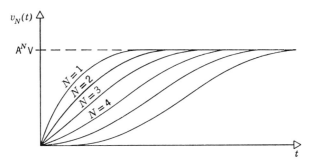

FIGURE 7-7. Output of cascaded amplifiers for a step function. Input for various numbers of stages.

From this set of curves we can see that the effect of adding additional stages of amplification is to introduce delay into the circuitry and to increase the effective rise time of the overall amplifier. Of course, the advantage is the increase in gain.

Elmore* has shown that for cascaded amplifiers with monotonic outputs, that is, no ringing, general relationships may be derived from the transfer functions which relate the overall characteristics of the amplifier to those of the individual stages. These are

$$A = A_1 \cdot A_2 \cdots A_N \qquad \text{gain}$$

$$T = T_{d_1} + T_{d_2} + \cdots + T_{d_N} \qquad \text{delay}$$

$$T_R^2 = T_{R_1}^2 + T_{R_2}^2 + \cdots + T_{R_N}^2 \qquad \text{rise time}$$

We note that if all stages are equivalent, these equations reduce to

$$A = A_1^N;$$

$$T = NT_{d_1};$$

$$T_R = \sqrt{N}\, T_{R_1}.$$

GAUSSIAN AMPLIFIER

It is of interest to consider the limiting case of an infinite number of stages in order to see the general shape of the frequency response of a cascaded amplifier. We obtain the response of the system to a sinusoidal input by replacing s

* W. C. Elmore, *J. Appl. Phys.* **19**, 55 (1948).

AMPLIFIERS WITH ACTIVE ELEMENTS

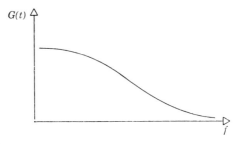

FIGURE 7-8. Frequency response of a Gaussian amplifier.

by $j\omega$ and rationalizing the result. Hence,

$$\frac{T_N(\omega)}{A^N} = (1 + \omega^2\tau^2)^{-N/2} e^{-jN \tan^{-1} \omega\tau}.$$

The frequency response is then given by

$$G(\omega) = (1 + \omega^2\tau^2)^{-N/2}.$$

We may expand this into a binomial series to give

$$G(\omega) = 1 - \frac{N\tau^2\omega^2}{2} + \frac{N}{2}\left(\frac{N}{2} + 1\right)\frac{\tau^4\omega^4}{2!} - \cdots$$

and take the limit as $N \to \infty$. Thus,

$$\lim_{N \to \infty} G(\omega) = 1 - \frac{N\tau^2}{1!}\left(\frac{\omega^2}{2}\right) + \frac{(N\tau^2)^2}{2!}\left(\frac{\omega^2}{2}\right)^2 - \cdots.$$

But if all of the stages are equivalent, we have

$$T_R = \sqrt{N}\, 2.2\tau \simeq \sqrt{N2\pi}\,\tau.$$

Hence,

$$G(f) \simeq 1 - \frac{\pi T_R^2 f^2}{1!} + \frac{(\pi T_R^2 f^2)^2}{2!} - \cdots,$$

or

$$G(f) \simeq e^{-\pi T_R^2 f^2}.$$

Thus, the frequency response is Gaussian (Fig. 7-8).
The upper half-power point (bandwidth in this case) is found from the condition

$$e^{-\pi T_R^2 f^2} = \frac{1}{\sqrt{2}}.$$

EXPERIMENTS 9 AND 10

Thus,
$$T_R f_2 = \sqrt{\frac{\ln 2}{2\pi}},$$
or
$$T_R f_2 \simeq \tfrac{1}{3}.$$

The latter relationship is extremely useful for relating bandwidth to rise time.

Experiments 9 and 10

By this point, we feel that the student should be able to design and construct both the Darlington circuit and a difference amplifier without too much trouble. These circuits should be well understood as far as their ac operation is concerned. The dc circuits should be designed to place the transistors or FETs in their active regions. Before doing this, be sure that the biasing circuit does not interfere with the operation of the ac circuit. For example, placing low-value biasing resistors across the Darlington input would drop the input impedance of the pair.

Once the circuits are operating, ask yourself what parameters are of the greatest interest in each circuit. Certainly, the input and output impedances and gain of a Darlington pair are of great interest. For the difference amplifier, input and output impedances are valuable, but gain and common-mode rejection ratio are of primary importance. These parameters can, of course, be measured as a function of frequency. Good luck.

OPERATIONAL AMPLIFIERS*

THE OP-AMP CONFIGURATION

We now wish to discuss one of the most important techniques of circuit analysis and design available today. With proper feedback a high-gain amplifier circuit can be arranged so that the transfer function of the circuit depends solely upon the circuit components *external* to the high-gain amplifier. This possibility immediately gives us great flexibility in designing functional circuits. Originally such circuits were used to perform mathematical operations for analog computers and were called *operational amplifier circuits*.

We will now consider single-ended op amps and derive the general form of the voltage transfer function $\bar{T}(s)$. In addition, we will derive and discuss the extremely useful concept of *virtual ground*.

---

* A good reference is *Handbook of Operational Amplifier Applications*, Burr-Brown Corp., Tucson, Arizona (1963). See also J. G. Graeme, G. E. Tobey and L. P. Huelsman, *Operational Amplifiers, Design and Applications* (*Burr-Brown*). McGraw-Hill, New York (1971).

OPERATIONAL AMPLIFIERS

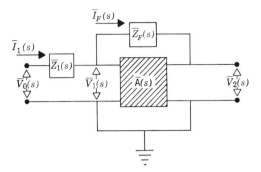

FIGURE 8-1. Operational-amplifier configuration.

Let us consider the fedback voltage amplifier of Fig. 8-1. We will now make a number of assumptions about the unfedback amplifier:

1. Input impedance very high, $= \infty$.
2. Output impedance very low, $= 0$.
3. Gain very high (often $> 10^6$).
4. Infinite bandwidth.

Basically, we have a high-gain amplifier with the capability of amplifying dc signals.

In actual practice, however, the above conditions are only partially satisfied. For example, the bandwidth is always finite, and at high enough frequencies, the gain will fall off. Thus, the third assumption above will not be satisfied, and the operational amplifier will not perform as advertised for the high-frequency components of a given waveform (see Problems).

We will use the notation of Fig. 8-2, where the ground terminal leads have been suppressed. The minus sign is due to sign reversal in most voltage amplifiers and is sometimes omitted. Let us apply Kirchhoff's laws to the circuit shown in Fig. 8-1:

$$\bar{V}_0 - \bar{I}_1 \bar{Z}_1 - \bar{I}_F \bar{Z}_F - \bar{V}_2 = 0;$$

$\bar{I}_1 = \bar{I}_F = \bar{I}$ (∞ input impedance for the unfedback amplifier);

$$\bar{V}_0 - \bar{I}\bar{Z}_1 - \bar{V}_1 = 0;$$
$$\bar{V}_2 = -\bar{A}\bar{V}_1.$$

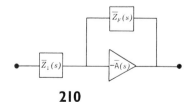

FIGURE 8-2. Simplified schematic of an Op-amp configuration.

CONCEPT OF THE VIRTUAL GROUND

Solving these equations, we have

$$\bar{I} = \frac{\bar{V}_0 - \bar{V}_1}{Z_1}.$$

Then

$$\bar{V}_2 = \bar{V}_0 - (Z_1 + Z_F)\left(\frac{\bar{V}_0 - \bar{V}_1}{Z_1}\right),$$

or

$$\bar{V}_2 = \bar{V}_0 - \frac{(Z_1 + Z_F)}{(-\bar{A})Z_1}(-\bar{A}\bar{V}_0 - \bar{V}_2);$$

thus,

$$\frac{\bar{V}_2}{\bar{V}_0} = \bar{T}(s) = -\frac{Z_F}{Z_1}\left[\frac{1}{1 + \frac{1}{\bar{A}}\left(1 + \frac{Z_F}{Z_1}\right)}\right].$$

For

$$\left|\left(\frac{1}{\bar{A}}\left(1 + \frac{Z_F}{Z_1}\right)\right)\right| \ll 1,$$

we have

$$\bar{T}(s) = -\frac{Z_F(s)}{Z_1(s)},$$

the *Operational-Amplifier Transfer Function*, where we recall

Z_1 = operational input impedance;
Z_F = operational feedback impedance.

We have, then, the important result that the transfer function is *not* a function of the main amplifier characteristics as long as the gain of the main amplifier is high enough.

CONCEPT OF THE VIRTUAL GROUND

Let us now solve for $\bar{V}_1(s)$:

$$\bar{V}_2 = -\bar{A}\bar{V}_1 = \bar{V}_0 - (Z_1 + Z_F)\left(\frac{\bar{V}_0 - \bar{V}_1}{Z_1}\right);$$

$$\bar{V}_1\left[-\bar{A} - \left(1 + \frac{Z_F}{Z_1}\right)\right] = \bar{V}_0\left[1 - \left(1 + \frac{Z_F}{Z_1}\right)\right];$$

$$\bar{V}_1 = \frac{-\frac{Z_F}{Z_1}}{-\bar{A} - \left(1 + \frac{Z_F}{Z_1}\right)}\bar{V}_0,$$

OPERATIONAL AMPLIFIERS

or

$$\bar{V}_1 = \frac{1}{1 + \frac{Z_1}{Z_F}(1 + \bar{A})} \bar{V}_0 \simeq 0$$

if

$$\left|\frac{Z_1}{Z_F}(1 + \bar{A})\right| \gg 1, \text{ since } A \gg 1.$$

Therefore, $|\bar{V}_1| \ll |\bar{V}_0|$. Also, we have $\bar{V}_1 = \bar{V}_2/(-\bar{A}) \simeq 0$, so that $|\bar{V}_1| \ll |\bar{V}_2|$. Hence, $\bar{V}_1$ is at *virtual ground*, virtual because no current flows through it to ground.

We note again that the *input impedance* of the op amp, taking into consideration the virtual ground, is just $\bar{Z}_1$. The output impedance will be just that of the high-gain amplifier which is very close to zero (see Problem 8-3). Voltage buffering will thus be quite easy with op amps.

OPERATIONAL CONFIGURATIONS FOR DIFFERENTIATION, INTEGRATION, ETC.

We now consider a number of op-amp circuits which may be easily analyzed. The simplest circuit is shown in Fig. 8-3. Here $\bar{T}(s) = -R_F/R_1$, so the circuit acts as an *inverter amplifier* with gain $A = R_F/R_1$. Of course, a simple *inverter* is obtained if $R_F = R_1$.

The operational amplifier makes it possible to obtain "perfect" differentiation and integration. For example, the op-amp differentiator is shown in Fig. 8-4.

The transfer function is

$$\bar{T}(s) = -\frac{R}{\frac{1}{sC}} = -s\tau.$$

The response to an arbitrary voltage input $v_6(t)$ is found from

$$\bar{V}_2(s) = -\tau s \bar{V}_0(s),$$

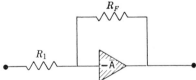

FIGURE 8-3. Op-amp inverter amplifier.

OP-AMP CONFIGURATIONS

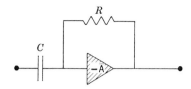

FIGURE 8-4. Op-amp differentiator.

whose inverse transform is

$$v_2(t) = -\frac{\tau \, dv_0}{dt}.$$

This form of integrating circuit is to be contrasted with the simple RC differentiator circuit whose transfer function was found earlier to be

$$\bar{T}(s) = \frac{s\tau}{1 + s\tau},$$

which looks like a differentiator only for its long-term (or low-frequency) response.

Let us now consider the *Miller integrator op-amp circuit* (Fig. 8-5). The transfer function of this circuit is

$$\bar{T}(s) = \frac{-\dfrac{1}{sC}}{R} = -\frac{1}{s\tau},$$

and its response to an arbitrary input $v_0(t)$ is obtained from

$$\bar{V}_2(s) = -\frac{1}{s\tau} \bar{V}_0(s),$$

whose inverse transform is

$$v_2(t) = -\frac{1}{\tau} \int_0^t v_0(t') \, dt'.$$

This form of integrating circuit is to be contrasted with the simple RC integrator whose transfer function was found earlier to be

$$\bar{T}(s) = \frac{1}{1 + s\tau},$$

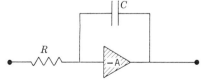

FIGURE 8-5. Op-amp integrator.

213

OPERATIONAL AMPLIFIERS

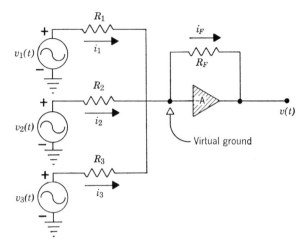

FIGURE 8-6. Op-amp adder.

which looks like an integrator only for its short-term (or high-frequency) response.

As an example of the use of the virtual ground concept, let us construct the *summing*, or *adder*, op-amp circuit shown in Fig. 8-6.

We write a nodal equation at the virtual ground point,

$$\bar{I}_1 + \bar{I}_2 + \bar{I}_3 - \bar{I}_F = 0,$$

and neglect, as before, the current input to the high-gain amplifier. Now, since we have a virtual ground at the summing point, we may write individual voltage loop equations:

$$\bar{V}_1(s) - \bar{I}_1 R_1 = 0;$$
$$\bar{V}_2(s) - \bar{I}_2 R_2 = 0;$$
$$\bar{V}_3(s) - \bar{I}_3 R_3 = 0;$$
$$\bar{V}(s) + \bar{I}_F R_F = 0.$$

If we make $R_1 = R_2 = R_3 = R$, we have, for the Laplace transform of the output voltage,

$$\bar{V}(s) = -\frac{R_F}{R}(\bar{V}_1 + \bar{V}_2 + \bar{V}_3),$$

whose inverse transform is

$$v(t) = -\frac{R_F}{R}[v_1(t) + v_2(t) + v_3(t)],$$

and we have a summing inverter amplifier.

A CHARGE-SENSITIVE PREAMPLIFIER

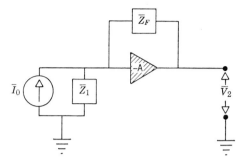

FIGURE 8-7. Op-amp preamplifier.

A CHARGE-SENSITIVE PREAMPLIFIER

We now point out an interesting property of op amps driven by current sources and illustrate it with a very useful circuit called a *charge-sensitive preamp*. Suppose we have the circuit shown in Fig. 8-7. We may replace the Norton current driver by its Thevenin voltage equivalent to give the circuit of Fig. 8-8.

We now obtain the output voltage by using the op-amp transfer function. Hence,

$$\bar{V}_2(s) = -\frac{\bar{Z}_F}{\bar{Z}_1} \bar{I}_0 \bar{Z}_1,$$

or

$$\bar{V}_2(s) = -\bar{I}_0 \bar{Z}_F.$$

Note: The reader may wish to verify this result by using Kirchhoff's laws in Fig. 8-7 and the assumption of high gain for the unfedback amplifier.

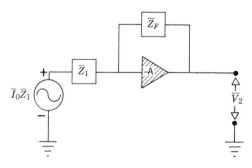

FIGURE 8-8. Equivalent circuit for an op-amp preamplifier.

OPERATIONAL AMPLIFIERS

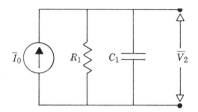

FIGURE 8-9. Charge-to-voltage converter.

Note that Z_1 does not appear in the expression for the output voltage. The charge-sensitive preamplifier illustrates some of the advantages of using an operational amplifier. There are many cases in which one wants a voltage output proportional to total charge, that is, the integral of the current source. Normally, one just uses a simple parallel RC integrating circuit, as, for example, with photomultiplier outputs (Fig. 8-9). Here the output is

$$\bar{V}_2(s) = I_0 \bar{Z}_1,$$

where

$$\bar{Z}_1 = \frac{1}{C_1}\left(\frac{1}{s + (1/\tau_1)}\right)$$

and

$$\tau_1 = R_1 C_1.$$

This is quite satisfactory as long as R_1 and C_1 remain constant. However, if one is using a device such as a semiconducting particle detector, the effective capacitance will be voltage dependent. The charge-sensitive preamp allows one to effectively replace the variable capacitance C_1 by a stable capacitor C_F in the feedback loop and at the same time adjust the time constant to an optimum value (Fig. 8-10).

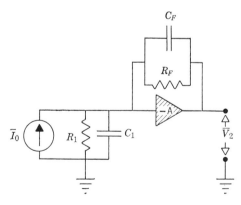

FIGURE 8-10. Charge-sensitive preamplifier.

AN ACTIVE DELAY LINE DIFFERENTIATOR

The output voltage $\bar{V}_2$ is related to the input current $\bar{I}_0$ by the *transfer impedance* $\bar{Z}_T = -\bar{Z}_F$, since $\bar{V}_2(s) = \bar{I}_0 \bar{Z}_T = \bar{I}_0(-\bar{Z}_F)$. Now

$$\bar{Z}_F(s) = \frac{1}{C_F}\left(\frac{1}{s + (1/\tau_F)}\right)$$

and

$$\tau_F = R_F C_F.$$

We still have not shown that the charge-sensitive preamp is sensitive to charge. Assume that R_F is very large. Then

$$\bar{Z}_F(s) \simeq \frac{1}{sC_F}$$

and

$$\bar{V}_2(s) \simeq -\frac{\bar{I}_0(s)}{sC_F}.$$

Recalling the Laplace transform of the indefinite integral, we can write

$$v_2(t) \simeq -\frac{1}{C_F}\int_0^t i_0(t')\,dt';$$

$$v_2(t) \simeq -\frac{q_0(t)}{C_F}.$$

AN ACTIVE DELAY-LINE DIFFERENTIATOR

As another example of the use of operational amplifiers, let us consider the case of the *active delay-line differentiator*. It will be recalled that in Chapter 3 a shorted delay line, or delay-line clip, was used to generate a narrow voltage pulse. There are two disadvantages to such a circuit which have not been previously discussed. The first is that the delay line is impedance matched only at one end. Since it is difficult to obtain perfect matching, there will be some unwanted reflections appearing in the circuit output. The second difficulty is that when attenuation is taken into account, there is not perfect cancellation of the forward and backward waves, resulting in a "rear porch," or tail, which may again lead to pile-up for high pulse-counting rates (Fig. 8-11).

These objections are removed in the active delay-line differentiator shown in Fig. 8-12.

We would first like to terminate the delay in its characteristic impedance at *both* ends. This we can do very nicely by making use of the properties of op amps. For the front of the line, we want to drive the line through the impedance Z_0 with an ideal voltage source. The low output impedance of the

OPERATIONAL AMPLIFIERS

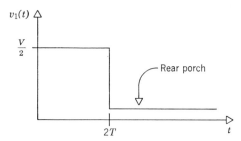

FIGURE 8-11. Output of a delay-line clip.

op-amp makes it an ideal candidate for this job. At the end of the line, we may terminate with an op amp of input impedance Z_0 since the resistance Z_0 will be connected to virtual ground. In order to achieve the desired pulse shaping, we must subtract the delayed output from the original signal. This can be accomplished by inverting the delay output with an op-amp inverter and adding the result algebraically to the original voltage in a summing op amp. By making the driving op amp an inverter amplifier with adjustable gain, we may eliminate the objectional "rear porch" effect due to attenuation in the line.

The output voltage then becomes

$$\bar{V}_2(s) = -\left[\frac{R_F}{R_2}\bar{V}_0 + \frac{R_F}{Z_0}e^{-\alpha l}\frac{e^{-sT}}{2}\left(-\frac{R_{F1}}{R_1}\right)\bar{V}_0\right],$$

or

$$\bar{V}_2(s) = -\frac{R_F}{R_2}\left[1 - \left(\frac{R_2}{Z_0}\right)\frac{R_{F1}}{R_1}\left(\frac{e^{-\alpha l}}{2}\right)e^{-sT}\right]\bar{V}_0(s).$$

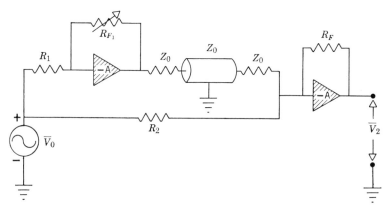

FIGURE 8-12. Active delay-line differentiator.

LINEAR INTEGRATED CIRCUITS

We now wish to adjust the factor in front of e^{-sT} to be equal to one so that the resulting output is a true rectangular pulse, with no rear porch, for a step voltage input:

$$\bar{V}_2(s) = -\frac{R_F}{R_2} V_0 \left(\frac{1 - e^{-sT}}{s} \right).$$

There is still flexibility in the choice of the parameters R_2 and R_{F_1}/R_1, but reasonable values would be $R_2 = Z_0/2$ and $R_{F_1}/R_1 > 4$.

LINEAR INTEGRATED CIRCUITS

The applications possible for operational-amplifier configurations are virtually unlimited, and their utilization would have greatly increased even without the advent of a device ideally suited to the concept, the linear integrated circuit. However, with the advent of the linear integrated circuit, it becomes possible to mass produce the high-gain dc-coupled amplifiers that are the heart of the operational amplifier and yet to use no more space than a transistor. The technique is rapidly advancing into large-scale integration (LSI), in which whole systems are contained on a single chip.

A linear integrated circuit consists of arrays of transistors, resistors, and capacitors formed into a single semiconducting crystal and performing the same tasks as equivalent circuits consisting of discrete components. The important point is that most of the size of a transistor is generated by mechanical considerations, such as having to support the crystal, and thus most of the semiconducting crystal is wasted. Therefore, considerably greater utilization of the crystal can be achieved while not increasing the size of the package.

Linear integrated circuits can be made, using either bipolar transistors or FETs, depending on whether one uses the silicon planar technology of transistors or the thin film techniques of MOS devices. Regardless of the precise technique used to form the integrated circuit, there is a relationship between the element chosen and the relative area of the chip required for the element. For silicon planar technology,*

Element	Relative Area
Transistor	1
Resistor, 1 K	2
Capacitor, 10 pF	3

* *RCA Linear Integrated Circuits Handbook*, Radio Corporation of America, New York (1967).

OPERATIONAL AMPLIFIERS

Therefore, the circuits that best utilize the integrated-circuit concept will have more active elements than passive elements, especially capacitors. This is precisely the inverse of what a circuit designed with discrete components tends to do.

It is difficult to form resistors that have a precise value, while the relative value of two resistors can be very accurate. Also, since the transistors are formed on the same chip and are close to each other, their relative parameters are well matched, and temperature effects should be the same for the pair.

Two circuits that fulfill these requirements very well are the difference amplifier and operational amplifiers based on difference amplifiers. A very important point is that the difference amplifier is extremely versatile, and by changing the *outboard components* external to the amplifier, one can form a large variety of circuits. The operational amplifier integrated circuit commonly uses a cascade of balanced differential amplifiers and an output stage. The critical feedback components are "outboarded" since usually the required stability and versatility can not be attained with integral components. For further detail of construction and utilization, including the avoidance of unwanted oscillations, one should consult the references given at the beginning of this chapter.

Finally, one should notice that the integrated circuit lends itself very well to the type of analysis used in this chapter. The basic amplifier $\bar{A}(s)$ is actually an integrated circuit, while $\bar{Z}_1(s)$ and $\bar{Z}_F(s)$ are discrete components. The details of the integrated circuit performance are not important as long as the gain is high enough.

LOGARITHMIC AMPLIFIERS

There are many situations where one would like to convert a given signal to the logarithm of that signal. For example, one may wish to have a direct reading of a decibel level, or it might be necessary to convert an optical transmission change to an absorbance. It is possible to do this in an analog fashion by making use of the exponential relationship between voltage and current in a forward-biased *pn* junction.

In Fig. 8-13 we show one possible circuit for stable logarithmic conversion of a voltage signal.

The collector current in Q_1 is given approximately by the following relation:

$$I_{C1} = I_{01}(e^{V_{BE_1}/\eta V_T} - 1).$$

LOGARITHMIC AMPLIFIERS

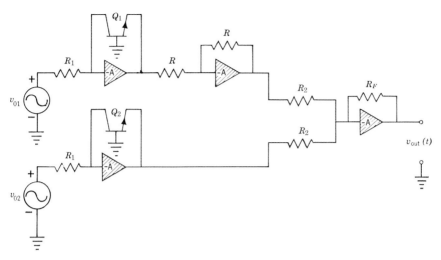

FIGURE 8-13. Logarithmic amplifier.

The -1 in the above equation may be neglected if the base-emitter junction is highly forward biased. We have, then, the following logarithmic relation:

$$V_{BE_1} = \eta V_T \ln \frac{I_{C_1}}{I_{01}},$$

where $V_T = RT/e$, as before. Using the virtual ground concept for the first op-amp, we see that

$$I_{C_1} \simeq \frac{v_{01}}{R_1}.$$

Thus,

$$V_{BE_1} = \eta V_T \ln \frac{v_{01}}{I_{01} R_1},$$

and we have the desired logarithmic relationship. However, the reverse saturation current I_{01} is a very sensitive function of temperature. With the circuit given above, we may cancel out this temperature sensitivity by using matched transistors and subtracting the outputs of the two log amps in a difference amplifier. In the above circuit, we have constructed an op-amp difference amplifier by combining an op-amp inverter and an op-amp summing circuit.

The output of the overall circuit is then given by

$$v_{out}(t) = -\frac{R_F}{R_2}(V_{BE_2} - V_{BE_1}),$$

OPERATIONAL AMPLIFIERS

or

$$v_{out}(t) = -\frac{\eta V_T R_F}{R_2}\left[\ln\left(\frac{v_{02}}{R_1 I_{02}}\right) - \ln\left(\frac{v_{01}}{R_1 I_{01}}\right)\right].$$

This may be written in the following form:

$$v_{out}(t) = -\frac{\eta V_T R_F}{R_2}\left(\ln v_{02} - \ln v_{01} + \ln\frac{I_{01}}{I_{02}}\right).$$

We now see the great advantage of the above circuit. The last term will be approximately ln 1, or zero, if the transistors are matched and in the same environment; usually they are on the same integrated chip. They will respond to temperature variations in the same way, so that the ratio I_{01}/I_{02} will stay constant near a value of one. We then have directly

$$v_{out}(t) = K(\ln v_{02} - \ln v_{01})$$

If the photomultiplier output voltages v_{01} and v_{02} are proportional to optical transmissions under two different circumstances, the above circuit allows us to obtain the desired optical absorbance change in an analog fashion.

PROBLEMS

1. Derive the voltage transfer functions for the op-amp circuits as shown on p. 223 and at top of p. 224, and discuss their functional properties. Which parameter would you vary in order to change the time constant?
2. Discuss the response of an RC integrator circuit to an arbitrary voltage input by making use of the fact that the inverse Laplace transform of the product of two transforms is the convolution of the inverse transforms of the individual transforms. (Look up the convolution theorem in a Laplace transform textbook.) Over what interval would the RC integrator circuit be a good integrator?
3. Derive the output impedance for an operational-amplifier configuration.
4. As stated in this chapter, the amplifier that is the heart of an operational-amplifier configuration has the rather stringent requirements of having infinite input impedance, gain, and bandwidth and zero output impedance. In lieu of such perfection, let us suppose that the amplifier has an input impedance of 10^5 ohms, bridged by a stray capacitance of 10 pf, a gain of 10^5 with a -3-dB point of 10^5 Hz, and an output impedance of 100 ohms.
   (a) Design an amplifier with a gain of 10, using the ideal amplifier. Use as the input impedance 125 ohms.
   (b) Now use the "realistic" amplifier in your design. Describe what happens to the input impedance, output impedance, and gain of the amplifier as a function of the frequency of an input sine wave.

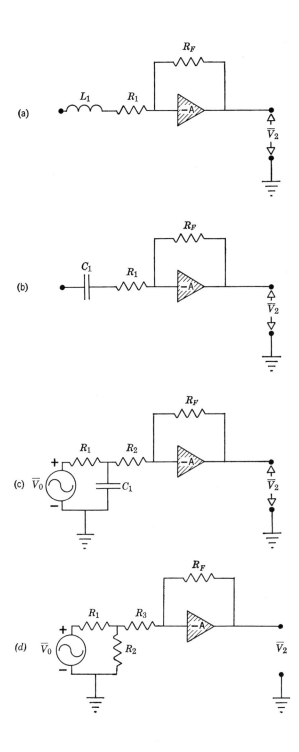

OPERATIONAL AMPLIFIERS

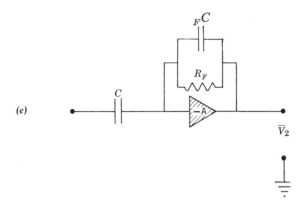

(e)

Special Project in Operational Amplifier Analysis

On the tip in, you will find block and schematic diagrams* of a linear amplifier designed to handle fast pulses. This is the first stage of the Oak Ridge Technical Enterprises Corporation's Model 220 Linear Pulse Analysis System, a very popular pulse analysis system in recent years in the field of nuclear electronics and other fields that require the handling of fast linear pulses.

The linear amplifier stage has to perform a number of tasks. The first of these is to amplify the incoming pulse by a variable factor that can range between 0.35 and 1300. No less important, the amplifier must shape the incoming pulse to maximize the informational content while minimizing the noise. This is usually done by deliberately decreasing the bandwidth of the amplifier by integrating and/or differentiating the incoming pulse for optimum signal-to-noise ratio (see Chapter 9). Finally, the system must protect itself from overvoltages within each stage as well as from overvoltages present at the input.

Ignore for the moment the diodes present in the circuit. They will be treated in Chapter 10.

Using drawing 220-101-B1, explain the purpose and operation of every section of the circuit, working from virtual ground to virtual ground. Break up each element, analyse its operation, and explain how it fits in with the whole system. Use values given in drawing 220-101-S1 when such are not indicated in the first diagram. Although not indicated, the input attenuator serves the purpose of changing the coarse gain in steps of 1, 2, 5, 10, 20, and 50 while maintaining a 125-ohm impedance at the input. Please indicate how this can be done.

* Diagrams reproduced with the permission of Oak Ridge Technical Enterprises, Inc.

EXPERIMENT II

Experiment 11
Operational Amplifiers and Feedback

This experiment is an introduction to integrated-circuit operational amplifiers, and a demonstration of the effect of feedback on amplifier characteristics. Both regenerative and degenerative feedback, as well as resistive and reactive feedback elements, will be used.

The central device is an RCA CA3000 integrated-circuit dc amplifier. The circuit diagram and terminal connections are shown.* Also note the tables giving maximum voltage limits and electrical characteristics, and the glossary of terms.

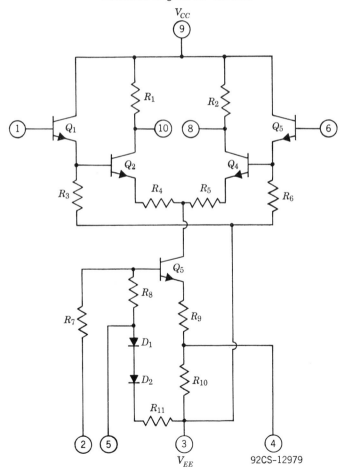

Schematic diagram for CA 3000

* Specification sheets reproduced with permission of the Radio Corporation of America.

OPERATIONAL AMPLIFIERS

The CA3000 circuit is basically a single-stage differential amplifier (Q_2 and Q_4) with input emitter-followers (Q_1 and Q_5) and a constant-current sink (Q_3) in the emitter-coupled leg. The use of degenerative resistors R_4 and R_5 in the emitter-coupled pair increases the linearity of the circuit and decreases its gain. The diodes D_1 and D_2 are for temperature stabilization.

Equipment

1. 1 FET VOM
2. 1 Dual power supply
3. 1 Signal (scan wave) generator, low output impedance
4. 1 Oscilloscope
5. 1 CA3000 I-C dc amplifier (RCA)
6. 2 0.5 µf to 1 µf capacitor
7. 1 100 pf capacitor
8. 1 Decade capacitor
9. 2 1 K-ohm to 2 K-ohm resistor
10. 1 8 K-ohm to 10 K-ohm resistor
11. 2 Decade resistors
12. 1 Decade inductor

Dimensional outline for CA3000

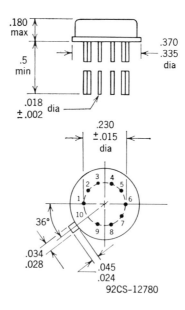

92CS-12780

Absolute-Maximum Voltage Limits, at $T_{FA} = 25°C$

TERMINAL	VOLTAGE LIMITS NEGATIVE	VOLTAGE LIMITS POSITIVE	CONDITIONS TERMINAL	CONDITIONS VOLTAGE
1	−2	+2	2 3 6 9	0 −6 0 +6
2	−8	0	1 3 6 9	0 −8 0 +6
3	−10	0	1 2 6 9	0 0 0 +6
4	−8	0	1 2 6 9	0 0 0 +6
5	−6	0	1 2 3 6 9	0 0 −6 0 +6
6	−2	+2	1 2 3 9	0 0 −6 +6
7	NO CONNECTION			
8	0	6	1 2 3 6	0 0 −6 0
9	0	+10	1 2 3 6	0 0 −6 0
10	0	+6	1 2 3 6	0 0 −6 0
CASE	Internally Connected to Terminal No. 3 (Substrate) DO NOT GROUND			

* Indicated voltage limits for each terminal can be used under specified voltage conditions for other terminals. All voltages are with respect to ground (common terminal of Positive and Negative dc Supplies)

Electrical Characteristics for CA3000, at $T_{FA} = 25°C$, $V_\infty = +6V$, $V_{EE} = -6V$, Unless Otherwise Specified

CHARACTERISTICS	SYMBOLS	SPECIAL TEST CONDITIONS Terminals No. 4 and No. 5 Not Connected Unless Specified		TEST CIRCUITS Fig.	LIMITS TYPE CA3000			Units	TYPICAL CHARAC- TERISTICS CURVES Fig.
					Min.	Typ.	Max.		
STATIC CHARACTERISTICS									
Input Offset Voltage	V_{10}			4	—	1.4	8	mV	2
Input Offset Current	I_{10}			5	—	1.2	10	μA	2
Input Bias Current	I_1			5	—	23	36	μA	3
		TERMINALS							
		4	5						
Quiescent Operating Voltage	V_8 or V_{11}	NC	NC	7	—	2.6	—	V	6
		NC	VEE	7	—	4.2	—	V	6
		VEE	NC	7	—	−1.5	—	V	6
		VEE	VEE	7	—	0.6	—	V	6
Device Dissipation	P_T	NC	NC	7	—	30	—	mW	NONE

DYNAMIC CHARACTERISTICS								
Differential Voltage Gain	A_{diff}	Single-Ended Output $f = 1$ kc/s	9	28	32	—	dB	8
Single-Ended Input		Double-Ended Output $f = 1$ kc/s	9	—	37	—	dB	8
Bandwidth at -3 dB Point	BW		11	—	650	—	kc/s	10
Maximum Output Voltage Swing	V_{out}(P-P)	$f = 1$ kc/s	9	—	6.4	—	V(P-P)	NONE
Common-Mode Rejection Ratio	CMR	$f = 1$ kc/s	13	80	98	—	dB	12
Single-Ended Input Impedance	Z_{in}	$f = 1$ kc/s	15	70 K	195 K	—	Ω	14
Single-Ended Output Impedance	Z_{out}	$f = 1$ kc/s	17	5.5 K	8 K	10.5 K	Ω	16
Total Harmonic Distortion	THD	$f = 1$ kc/s	19	—	0.2	5	%	18
AGC Range (Maximum Voltage Gain to Complete Cutoff	AGC	$f = 1$ kc/s	20	80	90	—	dB	NONE

OPERATIONAL AMPLIFIERS

DEFINITIONS OF TERMS FOR CA3000

Total Harmonic Distortion

The ratio of the total rms voltage of all harmonics to the rms voltage of the Fundamental, expressed in per cent. This voltage is measured at either output terminal with respect to ground.

Input Offset Voltage

The difference in the dc voltages which must be applied to the input terminals to obtain equal quiescent operating voltages (zero output offset voltage) at the output terminals.

Input Offset Current

The difference in the currents at the two input terminals.

Input Bias Current

The average value (one-half the sum) of the currents at the two input terminals.

Quiescent Operating Voltage

The dc voltage at either output terminal, with respect to ground.

DC Device Dissipation

The total power drain of the device with no signal applied and no external load current.

Common-Mode Voltage Gain

The ratio of the signal voltages developed between the two output terminals to the signal voltage applied to the two input terminals connected in parallel for ac.

EXPERIMENT II

Differential Voltage Gain—Single-Ended Input/Output

The ratio of the change in output voltage at either output terminal with respect to ground, to a change in input voltage at either input terminal with respect to ground.

Common-Mode Rejection Ratio

The ratio of the full differential voltage gain to the common-mode voltage gain.

AGC Range

The total change in voltage gain (from maximum gain to complete cutoff) which may be achieved by application of the specified range of dc voltage to the AGC input terminal of the device.

Bandwidth at −3-dB Point (BW)

The frequency at which the voltage gain of the device is 3 dB below the voltage gain at a specified lower frequency.

Maximum Output Voltage V_{out} (P–P)

The maximum peak-to-peak output-voltage swing, measured with respect to ground, which can be achieved without clipping of the signal waveform.

Single-Ended Input Impedance (Z_{in})

The ratio of the change in input voltage to the change in input current measured at either input terminal with respect to ground.

Single-Ended Output Impedance (Z_{out})

The ratio of the change in output voltage to the change in output current measured at either output terminal with respect to ground.

OPERATIONAL AMPLIFIERS

I. Circuit diagram

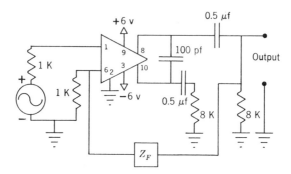

(Terminals on IC not shown above have no connection to them.)

II. Procedure

A. Build the circuit shown above.* For Z_F, use

1. Decade resistor
2. Decade capacitor

B. For each of 1. and 2. above, find and tabulate the gain and the bandwidth (to the 3-dB points) over the frequency range for various values of Z_F. Keep the input voltage small!

C. From a consideration of these results and past experience, what would you expect the gain-versus-frequency curve to look like for a purely inductive load? Verify by experiment if you're not sure.

D. In the case of resistive feedback, does the presence of feedback have any effect on the bandwidth of the amplifier?

E. Now for Z_F, use the following circuit and find the gain over the frequency range.

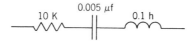

F. Compare the results of part E with your knowledge of LRC circuits, and use the results of the first parts of this experiment to explain the results of part E.

---

* For optimum operation, the quiescent voltages at the output terminals should be the same. You may wish to vary either of the input resistors until this condition is achieved. This will provide the correct *Input offset voltage*.

EXPERIMENT II

G. Now construct the following oscillator circuit. Explain the results, as in part F. Vary the values of the components and note the results. Include such observations as stability and shape of output waveform.

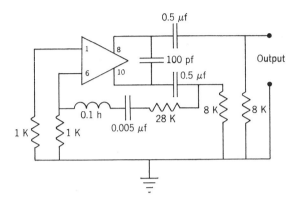

(All other connections are the same as before.)
Note: If your initial waveform is "imperfect," vary the 28K resistor.

NOISE 9

SOURCES OF NOISE

If we attempt to observe very small signals, we are immediately faced with the ever present problem of noise. Although the word is often used to denote all unwanted signals, we should restrict the use of the term to those cases that involve random fluctuations of voltage or current. In this chapter, we would like to briefly point out the sources of some of these unwanted signals and the methods that can be used to minimize their effects.

The most ubiquitous unwanted signals in the laboratory are what is often called *60 cycle*, that is, the 60-hertz powerline frequency and its harmonics. Electromagnetic radiation from the numerous powerline sources is easily picked up and seen by a high-gain amplifier system. There are several techniques to reduce this unwanted "pick-up." One method

NOISE

involves the use of grounded shields around the signal cables and apparatus. The wide use of singly and multiply shielded cables and enclosed metal chassis are an attempt to overcome this problem. Another is to make sure that the system is grounded at only one point in order to prevent *ground loops* in which unwanted signals can be generated by electromagnetic induction in a closed loop.

Another reason for eliminating ground loops is to prevent the situation that occurs if the two grounds are at slightly different potentials. Since shielding is designed to have low resistance, large currents can flow from ground to ground! This large current generates its own noise voltage by the techniques mentioned below. Finally, it should be noted that laboratory architects are always placing massive copper grounding busses from the upper floors of a building deep into the earth. Such a ground is adequate for dc work, but at high frequencies it may well act as a resonant antenna rather than as a ground. There is really no solution to this last problem except to be on the lower floors and not to mix high- and low-power high-frequency electronics in the same grounding system.

Nevertheless, such exterior sources of unwanted signals can be suppressed by a number of techniques. One is to use a symmetrical system with a difference amplifier for the input stage. Signals of this type will be picked up equally in both arms of the system and can be effectively eliminated by subtraction. One can also try to find frequency ranges in which the unwanted signals are at a minimum, and by using a limited, or narrow, band-pass system, reject signals except those in this narrow range. Problems with this last technique will be discussed later.

Theoretically, by the use of some or all of the above techniques, it should be possible to eliminate almost totally pick-up of the types discussed above. But there is a different type of unwanted signal which is more properly given the name noise. We are referring to voltage and current sources arising from the fact that actually nature is discrete, not continuous, and we are dealing with random variables which must be averaged to obtain our normal variables. Therefore, although we can obtain an average, we must also expect fluctuations of a purely statistical nature.

The first type of noise to be considered is Johnson, or thermal, noise. If we have a resistor at a temperature T, its atoms will be in a state of continual motion. This motion is partitioned among the electrons, thus presenting a fluctuating current. This fluctuating current in the resistor results in a fluctuating voltage, or noise. Since there is complete randomness, the average value across the resistor is zero, if we wait long enough; but there will be a finite value for the fluctuations. We may define an r.m.s. voltage such that

$$v_{\text{rms}}^2 = \overline{v^2(t)} \propto P;$$

SOURCES OF NOISE

that is, the time average of the square of the noise voltage is proportional to the noise power P. Since the source of energy for these voltage fluctuations is the vibrational motion of the resistive medium at temperature T, the noise power should be proportional to the vibrational energy, or the energy of the phonons, E.

If we are to minimize the effect of the thermal noise, we must know something about its frequency spectrum. Since the quantized vibrational energy or phonons follow Bose-Einstein statistics, the energy in a given frequency range df is given by

$$E = \frac{hf}{(e^{(hf/kT)} - 1)} df,$$

where h is Planck's constant and k is Boltzmann's constant. Hence, the average noise voltage squared in a given frequency interval df will be proportional to the above energy:

$$P = \frac{\overline{v^2}}{R} \propto \frac{hf}{(e^{(hf/kT)} - 1)} df.$$

For the frequencies normally encountered in electronics and at room temperature, we have the condition that

$$hf/kT \ll 1.$$

Therefore,

$$E = \frac{hf}{e^{(hf/kT)} - 1} df \simeq hf\left(\frac{1}{1 - hf/kT + \cdots - 1}\right) df.$$

Thus,

$$E \simeq kT \, df$$

and

$$\frac{\overline{v^2}}{R} \propto kT \, df.$$

A more rigorous analysis leads to

$$\overline{v^2} = 4kTR \, df;$$

that is, the frequency spectrum is flat, or the same for all frequencies. We speak, thus, of *white noise*. Since the noise voltage squared is proportional to the temperature, one obvious way of reducing the noise is to cool the resistor. This is the reason that top quality FET preamplifiers are operated at the temperature of liquid nitrogen, since R is so large in this device.

The second fundamental source of noise is called *shot noise* and results from the many cases in electronics in which electrons are emitted spontaneously from some surface or randomly cross some barrier, as in a transistor. If the emission of electrons is assumed to be random, then Poisson's

NOISE

distribution holds. An analysis of this statistical process* leads to the result that

$$i^2 = 2eI\,df,$$

where e is the electronic charge and I is the average current involved in the emission process.

The important result here is that both thermal noise and shot noise have a white, or frequency-independent, spectrum. Thus, we cannot get rid of the noise by moving our signal to a particular frequency. However, the fact that the noise power is proportional to the bandwidth df is most important.

A third source of noise that differs qualitatively from Johnson and shot noise was originally known as the "Flicker Effect" and is now known as *one-over-f noise* because of its inverse frequency dependence. It was originally noted that the noise in a vacuum tube, and later in transistors, was not flat at the lower frequencies but increased as the frequency of the noise decreased. The explanation that is most commonly used today is that this noise arises from rearrangements of the surface or interior structure of the substance, and these rather slow changes produce fluctuations that, when passed on to the electrons, produce a noise source with the observed $1/f$ frequency dependence. Detailed calculations of the effect are very difficult and not totally convincing, but improvements in materials and transistor design have reduced $1/f$ noise from a major, if not dominant, noise at almost all frequencies to a negligible effect except at very low frequencies, certainly well below 1 kHz. We might mention at this time that in a transistor at the high-frequency end of the spectrum, the effective noise also increases above the flat Johnson- and shot-noise values. This is not due to an increase in the noise as such, but due to the fact that α (and thus β) are decreasing as the α cut-off frequency is approached, so the ratio of signal to noise appears to be getting worse. There are certainly other sources of noise with other names, but basically all can be traced to mechanisms similar to the ones that we have discussed.

SIGNAL-TO-NOISE RATIO

From our discussion of the sources of noise, it is obvious that noise is here to stay, since the basic mechanisms are inherent in any electronic device. Therefore, the important concept is how much noise is present, not in an absolute sense, but in comparison to the information that we are trying to transmit. Since the instantaneous values of the noise voltage fluctuations are random and the average value zero, we must work with the average of the square of

---

* S. O. Rice in *Selected Papers on Noise and Stochastic Processes*, edited by N. Wax, Dover Publications, New York (1954), p. 133.

BANDWIDTH AND NOISE

the noise voltage. Therefore, at a given point in a system, we can define the signal-to-noise ratio as

$$\frac{S}{N} = \frac{v_s^2}{v_N^2}.$$

Therefore, the measurement of the noisiness of a system of elements can be expressed as the ratio of the signal to noise at the output to that at the input, or (where the input is v_0 and the output v_2)

$$\left.\frac{S}{N}\right|_2 = \frac{1}{n}\left.\frac{S}{N}\right|_0.$$

The parameter n is now a measure of the noisiness of the system and is called the *noise figure* of the circuit. Ideally, $n = 1$. Noise figures are frequently expressed in decibels, since the ratio involves two powers. Using the same definition as before,

$$\text{Noise}_{\#\text{dB}} = 10 \log_{10} n.$$

Therefore, if $n = 2$, the noise is $10 \log_{10} 2 = 10(0.3010) = 3$ dB. Often, total noise, and not just the increase in noise, is expressed in decibels, so that if $S/N = 100$, the noise is "down by 20 dB."

BANDWIDTH AND NOISE

To see the effect of bandwidth upon the fidelity of signal transmission through a circuit, let us consider the Fourier expansion of an infinite train of square-wave pulses (Fig. 9-1). It is a most important mathematical result that all periodic functions of physical interest may be expanded in terms of a series of sines and cosines.* This Fourier series may be written in the general form

$$f(x) = \frac{a_0}{2} + \sum_{n=1}^{\infty}(a_n \cos nx + b_n \sin nx)$$

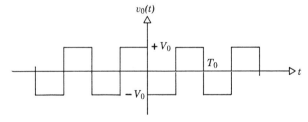

FIGURE 9-1. Infinite train of square-wave pulses.

---

* See I. S. Sokolnikoff and R. M. Redheffer, *Mathematics of Physics and Modern Engineering*, McGraw-Hill (1966).

NOISE

with

$$a_n = \frac{1}{\pi} \int_0^{2\pi} f(x) \cos nx \, dx;$$

$$b_n = \frac{1}{\pi} \int_0^{2\pi} f(x) \sin nx \, dx.$$

In our case, $x = (2\pi/T_0)t$, and the result is that

$$v_0(t) = \frac{-4V_0}{\pi}\left[\sin \omega t + \frac{\sin 3\omega t}{3} + \frac{\sin 5\omega t}{5} + \cdots\right]$$

and $\omega = 2\pi/T_0$. One can easily show that it is the higher-frequency components that generate the edges of pulses and any other sharp discontinuities, so that in the case of a square wave, an infinite number of frequencies must be retained to reconstruct the original signal. Thus, if one is working with very fast signals, a wideband amplifier is needed to handle the large range of frequencies Δf. However, this wideband amplifier will also amplify all the noise in Δf, and a point will be reached when any further increase in fidelity gained by increasing Δf is offset by the increased noise in the system. Realistic estimates must be made of the fidelity necessary for the work, and the amplifier should be matched to this optimum Δf. The usual practice is not to vary the amplifier but to precede the amplifier by an integrating and a differentiating circuit with controllable dominant time constants. A detailed analysis[*] shows that the optimum is reached when the two time constants are the same, or

$$\tau_{\text{diff}} = \tau_{\text{integ}}.$$

We can go farther, and now ask ourselves "What is the connection between the bandwidth of our system, the noise present, and the response time of the system to an input signal?" Assume that we wish to transmit information, and that we have a system that will respond with a characteristic rise time t_r ($= 2.2\tau$ for a simple integrating circuit). Assume further that we have a useful voltage range V and that we can distinguish two amplitudes that differ by ΔV. Therefore, the possible amplitudes, measured at t_r, or 90% of ultimate pulse height by our earlier definition, are just

$$\frac{V}{\Delta V} + 1 = a.$$

[*] A. B. Gillespie, *Signal, Noise and Resolution in Nuclear Counter Amplifiers*, Pergamon Press Ltd., New York (1954), p. 59.

BANDWIDTH AND NOISE

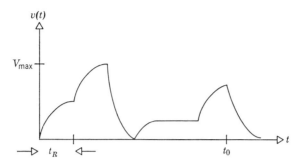

FIGURE 9-2. Train of pulses designed to transmit information.

Given an interval of time t_0 seconds long, we could then transmit a number of different combinations:

$$\# \text{ combinations} = a^{t_0/t_r}.$$

A sample combination, one of 15,625 for $a = 5$ (including 0 volts) and $t_0 = 6t_r$, is shown in Fig. 9-2. The voltage is thus sampled six times, equally spaced apart. Assuming zero noise, as we have done, we could put more information into this voltage interval, since the maximum amplitude variation is only about $0.1V_{max}$ while we have used a rather conservative $0.2V_{max}$ as our ΔV. Nevertheless, no tremendous improvement is possible, especially if there is uncertainty in the detection of these voltages. Doubling the time t_0 should simply double the information, so we shall define the relationship so this holds by using an exponential format. Therefore,

$$\text{information} = \text{number of combinations} = \frac{t_0}{t_r} \log_2 a.$$

The unit of information thus defined by our choice of a base-2 logarithm is called a *bit*. Now we can say that the maximum rate at which information can be transmitted is

$$C = \frac{\text{information}}{t_0} = \frac{1}{t_r} \log_2 a \text{ bits/second}.$$

We now need a more precise relationship between the bandwidth and the frequency response. One way to accomplish this is to return to our Fourier series and to rewrite our definitions in terms of exponentials instead of sines

241

NOISE

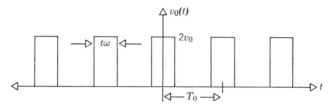

FIGURE 9-3. Infinite train of rectangular pulses.

and cosines. Trigonometric manipulation, using Euler's relations, gives

$$f(t) = \frac{1}{T_0} \sum_{n=-\infty}^{\infty} c_n e^{i\omega_n t};$$

$$c_n = \int_{-T_0/2}^{T_0/2} f(t) e^{-i\omega_n t} \, dt;$$

$$\omega_n = 2\pi n / T_0.$$

We will now apply this form to an infinite train of rectangular pulses whose period is, as before, T_0, and whose width is t_w (Fig. 9-3). We have chosen the voltage to be positive and the origin at the center of a pulse for convenience only. Then we can write

$$c_n = \int_{-t_w/2}^{t_w/2} 2V_0 e^{-i\omega_n t} \, dt;$$

$$c_n = \frac{4V_0}{\omega_n} \sin \frac{\omega_n t_w}{2};$$

$$c_n = 2V_0 t_w \frac{\sin y}{y},$$

where $y = \omega_n t_w / 2$. This function is plotted in Fig. 9-4.

We can use the fact that the power present as a function of frequency is proportional to the square of c_n, and this function is also plotted in Fig. 9-4. Most of the power, therefore, occurs in the central maximum, and we can use the first zero of the power spectrum for our definition of bandwidth. With this definition, we have that, for pulses much narrower than the period T_0,

$$\Delta f = \frac{1}{t_w}.$$

Although there is a degree of arbitrariness in our choice of Δf, the inverse relation with the pulse width will be maintained. This is exactly what we

BANDWIDTH AND NOISE

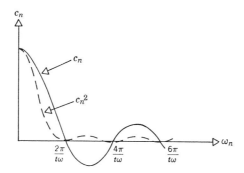

FIGURE 9-4. Amplitude and amplitude squared of the Fourier components of a train of rectangular pulses versus frequency.

meant before when we stated that measurements of fast signals required a large bandwidth and, therefore, much noise.

We can now rephase the maximum rate of transmission of information, using this definition of bandwidth. If we declare the pulse rise time t_r, to its 90% value from its 10% value, to be equal to t_w (recall the discussion of the Gaussian amplifier in Chapter 7), then

$$C = \frac{\text{information}}{t_0} = \Delta f \log_2 a.$$

But returning to our definition of the signal-to-noise ratio, we can see that the number of separable voltage levels a in the interval V_0 is approximately

$$a \simeq \left(1 + \frac{S}{N}\right).$$

Therefore, we obtain the important result

$$C \simeq \Delta f \log_2 \left(1 + \frac{S}{N}\right) \text{ bits/second.}$$

For the physically interesting cases in which the noise is proportional to the bandwidth (Johnson and shot noise), this relation reduces further to

$$C \simeq \Delta f \log_2 \left(1 + \frac{S}{K \Delta f}\right),$$

where K can be determined from the formulas presented in the first section of this chapter. As the bandwidth is decreased to very small values, the signal also suffers, so these approximate relations cannot be pushed too far in that

243

NOISE

direction. Nevertheless, the basic dependence of the transmission of digital information on the bandwidth and signal-to-noise ratio is exhibited above.

CAT AND LOCK-IN AMPLIFIERS

At first sight, we might think that once we have optimized our bandwidth and decreased our ambient temperature, we have done all we can to increase our signal-to-noise ratio with respect to thermal and shot noise. However, the fact that we are dealing with a random variable, whose average value is zero, means that if we can average, or integrate, our signals over sufficient time, we can average the thermal noise to zero. But we can't just stick a large integrating time constant into our circuit since this would cut our signal down to almost zero and certainly distort it. There are at least two ways we can get around this problem.

The first is the CAT method, that is, Computer Averaging over Time. If we have an exactly reproducible, recurring wave form, we may use an analog-to-digital converter and multichannel analyzer to algebraically add the recurring waveforms. The thermal noise voltage excursions will algebraically sum to zero, making it possible to pull out a signal deeply buried in thermal noise.

The second method is called the lock-in amplifier,* synchronous detection, or phase-sensitive amplifier technique of thermal- or shot-noise reduction. In this case, we are restricted to slowly varying signals, but the basic idea of averaging the thermal noise voltage to zero is still there. What one does is to introduce recurrent modulation into the original signal, perhaps by chopping it at some point, and derive from this a reference signal, usually

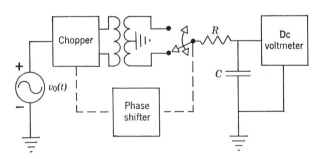

FIGURE 9-5. Simple version of a synchronous detector.

* J. C. Fisher, *TEK TALK* Vol. 6, No. 1, Princeton Applied Research Corporation.

CAT AND LOCK-IN AMPLIFIERS

a square-wave train, and then multiply these two signals together and integrate the result. If the phase between the chopped input signal and the reference signal is adjusted properly, that is, so that the reference signal is synchronous with the original chopped signal, then after mixing, a dc component proportional to the amplitude of the original signal will be present. Since we are dealing with slowly varying functions, the bandwidth necessary for good fidelity is fairly narrow to begin with, but this in itself presents problems if we modulate our signal and just try to use a narrow-bandwidth tuned amplifier. The problem is that, for really narrow bandwidths, the amplifier drifts would displace the bandwidth center from the signal frequency. For the case of the lock-in amplifier, the effective bandwidth center is "locked in" to the signal frequency.

The heart of the system is the synchronous detector, a simple version of which is shown in Fig. 9-5.

The circuit is seen to be, in essence, a full-wave voltage rectifier (see Chapter 10) and, hence, will yield a dc output proportional to the magnitude of the input signal. The thermal noise is then averaged to zero by the RC integrating time constant. The above scheme is also termed cross-correlation and is but one example of the very general and useful subject called correlation techniques.*

* L. Szmauz, "Correlation Function Computers," Princeton Applied Research Corporation Tech. Bull. No. 149 (1967).

NONLINEAR CIRCUIT ANALYSIS— DIODES

NONLINEAR SYSTEMS

In the first nine chapters, we have worked exclusively with linear elements and operations. Since all real devices are nonlinear to a greater or lesser extent, assumptions were made which "linearized" these systems, such as taking only the linear terms in a Taylor expansion of the voltage-current characteristics of a transistor or tube. We were then constrained to limit values of the input voltage or current so that these assumptions were fulfilled, and the result was a linear small-signal analysis. In the last two chapters, we are going to consider cases in which the voltage-current relationships are no longer even approximately linear, and we are going to try to handle these situations, using as much of our linear analysis as possible.

NONLINEAR CIRCUIT ANALYSIS—DIODES

One way to do this is to go to the other extreme and assume that di/dv is not only not a constant, but actually is infinite at a finite number of points! This is the *piecewise-continuous model*, in which the actual voltage-current relations are approximated by a finite number of linear regions separated by abrupt changes in the slope of the curve. Obviously, within each of the linear regions, all of our previous results are valid. This approach will lead us naturally to digital systems, as opposed to the analog linear systems, and we will develop in Chapters 10 and 11 basic circuitry for performing logical analysis of an input signal.

THERMIONIC AND JUNCTION DIODES

A diode is a two-terminal device which may be broadly characterized as having an effective resistance which is a function of the polarity of the input signal. This effective resistance will, in general, be much smaller in one sense of polarity than the other, resulting in two states of the system: the "on" state, in which the state conducts current freely, and the "off" state, in which the device conducts only very reluctantly.

There are many examples of devices that behave in this manner. One of the earliest examples was that of the vacuum, or thermionic, diode, a vacuum tube containing a heated cathode and a plate. Since the hot cathode is surrounded by a cloud of electrons, a positive voltage on the plate results in an immediate flow of electrons from the cathode to the plate, or an effective resistance that is relatively low. A negative voltage on the plate merely repels the electrons back towards the cathode, and very little current flows until the voltage on the plate rises high enough so that electrons are torn from the metal. This *cold-cathode emission* requires extremely high voltages in most well-designed vacuum diodes. At low voltages, the work function of the cathode and the initial velocity distribution of the electrons are the determining factors in the current flow, resulting in the relation

$$I = I_0 e^{eV/kT},$$

where I_0 is the value of the current flow for zero voltage V applied to the diode, and $kT = \frac{1}{40}$ electron-volt, the equivalent energy at room temperature of the electrons. Once the applied voltage is high enough so that the effects of the electron-cloud space charge are dominant, the relation becomes

$$I = KV^{3/2},$$

where the constant K is dependent upon the geometry used in the diode. Figure 10-1 shows the voltage-current relationship for a thermionic diode.

Gas-filled diodes, able to handle much higher currents than vacuum diodes, consist of thermionic diodes filled with some low-pressure gas, such

THERMIONIC AND JUNCTION DIODES

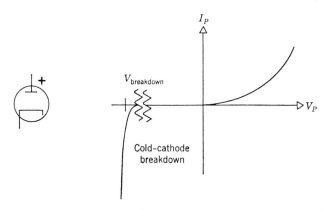

FIGURE 10-1. Schematic and characteristic curve for a thermionic diode.

as mercury vapor or a noble gas. They are commonly used for rectification in high-current applications. Finally, gas-filled diodes in which the cathode is not heated are used for voltage reference devices. Until the voltage is high enough to ionize the gas, little current flows. Once the ionization threshold has been reached, current carriers multiply and current flows easily. Most applications of these cold-cathode, gas-filled reference diodes have been superseded by Zener diodes, which act in a very similar manner.

Solid-state diodes have been around for a surprisingly long time, dating back to crystal radio sets in which a metal "whisker" was placed on a crystal to produce rectification of the input signal, an early form of the point-contact diode. For high-power applications, copper-oxide rectifiers, consisting of copper/copper-oxide/copper sandwiches, had the same result as modern diodes long before semiconductor physics became firmly established. These rectifiers have been largely replaced by selenium rectifiers, which can handle higher current densities and higher reverse voltages than the copper-oxide rectifiers. While point-contact diodes are still very much in evidence and are useful in many applications, we will discuss mainly the very common junction diode, formed of a single *pn* junction.

As we noted in Chapter 6, the current in an ideal junction diode is

$$I = I_0(e^{V/\eta V_T} - 1),$$

where $\eta = 1$ for Ge, and $\eta \simeq 2$ for Si; $V_T = 26$ mv, corresponding to the thermal energy at room temperature, and I_0 is in microamps for a Ge diode, nanoamps for an Si diode. Figure 10-2 shows the voltage-current relationship for a junction diode (silicon). Note that there is an effective voltage at which the diode starts to draw current in the range of 10's of milliamperes, the cut-in voltage V_c. This voltage is about 0.6 volts for silicon diodes and 0.2 volts

NONLINEAR CIRCUIT ANALYSIS—DIODES

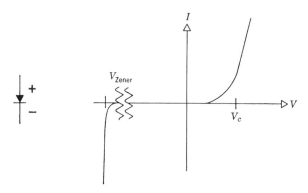

FIGURE 10-2. Schematic and characteristic curve for a junction diode.

for germanium diodes. At high enough reverse voltages, the diode again conducts abruptly at a Zener voltage V_Z due to a multiplication of the charge carriers. Reference, or Zener, diodes are designed to operate in this region; but for common diodes, operation in this region will result in rapid destruction of the device.

THE PIECEWISE-CONTINUOUS MODEL

As before, our first concern is to find the simplest model for the behavior of real diodes that yields adequate correlation with the observed effects. It should also be noted that the complexity of the model can be fitted to the task at hand, and often the most basic model will suffice in situations in which the exact waveform is not important to the result. We will approximate the characteristics of real diodes by the piecewise-continuous model, and compare the resulting "theoretical" waveforms with the actual waveform.

Consider the circuit of Fig. 10-3.

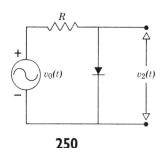

FIGURE 10-3. Clipping circuit.

THE PIECEWISE-CONTINUOUS MODEL

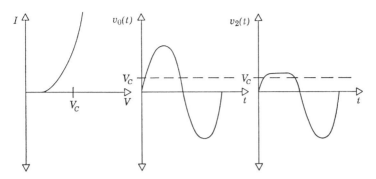

FIGURE 10-4. Characteristic curve, input, and output voltages for a junction-diode clipping circuit.

This simple circuit is extremely important in many applications. As you can see, a negative voltage across the diode (the diode is thus back biased) results in small current flow, and almost all the voltage at the input will appear at $v_2(t)$, while a positive input voltage will cause the diode to be forward biased (low effective resistance), and very little voltage will appear at $v_2(t)$. This circuit is called by many names, but very often the terms *clipping circuit* or *limiting circuit* are used, since it limits the excursions of the input in one sense of the voltage. Assuming that the input voltage is small enough so that the breakdown region at high negative bias is avoided, and that we have a silicon junction diode with characteristic curve as given in Fig. 10-2, the output will look like the curves shown in Fig. 10-4.

The simplest model is the piecewise-continuous model, in which the effective resistance of the back-biased diode is infinite while the forward resistance is zero for all voltages greater than zero. This would look like the

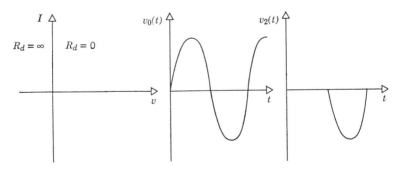

FIGURE 10-5. Characteristic curve, input, and output voltages for a clipping circuit—the first approximation.

NONLINEAR CIRCUIT ANALYSIS—DIODES

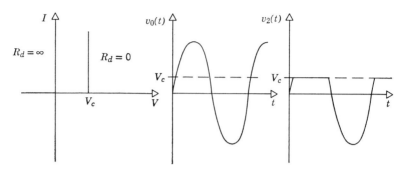

FIGURE 10-6. Characteristic curve, including V_c, input, and output voltages for a clipping circuit.

curves of Fig. 10-5. Although this model would suffice for many purposes, greater realism can be gained by introducing a cut-in voltage V_c, and the same analysis would become that shown in Fig. 10-6. This model now has one free parameter which can be adjusted, and V_c will depend on whether the diode is thermionic, germanium, or silicon. The prediction is certainly greatly improved.

One more step might be taken profitably. Consider a finite forward resistance (Fig. 10-7). The finite forward resistance means that the forward current flow does generate a small voltage-dependent output above V_c, which is observed in actuality. It is obvious that this procedure could be indefinitely extended to produce as close a fit as desired to reality by multiplying the break-points and resistances, but the law of diminishing returns is in evidence in the increasing number of parameters to be used.

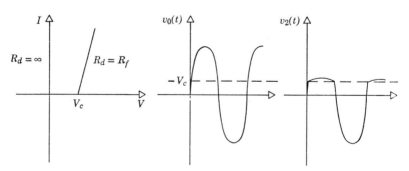

FIGURE 10-7. Characteristic curve, including V_c and finite forward resistance, input, and output voltages for a clipping circuit.

LIMITERS AND CLIPPERS

LIMITERS AND CLIPPERS

We can apply this model to various diode circuits by treating the diode as a resistor that has two values: $R_d = \infty$ below V_c, and $R_d = R_f$ above V_c. First, consider the circuit of Fig. 10-8. In this circuit, two diodes have been used as limiters to keep the level of the input signal below V_c in either the positive or negative sense. This is a simple, yet effective, method of protecting the circuits that follow. This type of circuit is adequate when one wishes to limit input to below V_c (0.2 to 0.6 volt approx.), but often signals larger than V_c are desired while limiting action is needed. This can be accomplished as shown below in Fig. 10-9.

Exactly the same waveform can be also generated by the circuit of Fig. 10-10. In our approximation, there is no preferred orientation, since the output is identical. However, in Fig. 10-9, the diode appears as a shunt element, while in Fig. 10-10 it appears as a series element. If we consider the effective capacitance of the diode, inherent in any *pn* junction (see Chapter 6), then it can be seen that when the diode is a shunt element, its effective resistance to ground will be less for high frequencies, and high frequencies will be attenuated. In the case of the series arrangement, high frequencies will tend to appear at $v_2(t)$ even when the diode is cut off for the same reasons. Therefore, a choice must be made as to which situation is the least damaging in any particular circumstance. Finally, clipping can be done at two independent levels by the use of two diodes and two references (Fig. 10-11).

We can arrange V_{R1} and V_{R2} to take any slice of the voltage that we wish. In particular, by having them equal and opposite, a sine wave can be made into a reasonable approximation of a square wave, the approximation becoming better as the peak voltage of the input is made much larger than the reference voltages. The limit is set by the breakdown voltages of the diodes and/or the capacitive effects of the two shunting diodes.

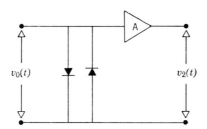

FIGURE 10-8. Voltage amplifier with limiters on the input.

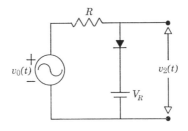

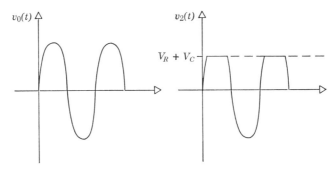

FIGURE 10-9. Variable bias limiter with input and output waveforms.

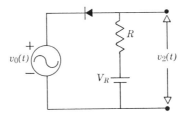

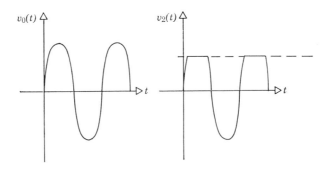

FIGURE 10-10. Variable bias limiter with input and output waveforms.

CLAMPING CIRCUITS

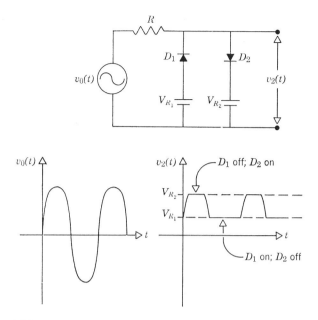

FIGURE 10-11. Clipping at two independent levels.

CLAMPING CIRCUITS

While these circuits scarcely exhaust the possibilities of diodes and resistive elements, we should also consider the addition of capacitors to diode circuits, which will allow further operations on the input waveforms. In particular, an important class of operations is labelled by the term *clamping*, generally signifying that some component or extremity of an ac waveform is forced to remain close to some reference voltage. Consider the circuit given in Fig. 10-12.

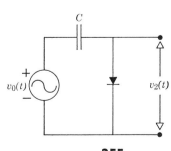

FIGURE 10-12. Clamping circuit.

NONLINEAR CIRCUIT ANALYSIS—DIODES

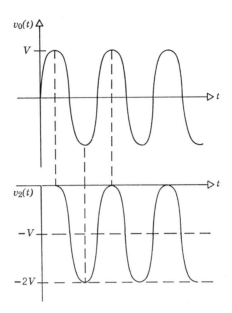

FIGURE 10-13. Input and output voltages of a clamping circuit.

While the input waveform rises, the diode conducts, and the only voltage that appears is the small voltage of the forward-biased diode. The capacitor charges up during this time. When the voltage drops, the diode is back biased, and the voltage of the input plus the voltage of the capacitor appears at the output (Fig. 10-13).

In this circuit, the voltage $+V_0$ is said to be clamped to ground. For obvious reasons, this circuit is often referred to as a *voltage-doubler circuit*. Note that the output $v_2(t)$ has a dc component lacking in the input. This leads

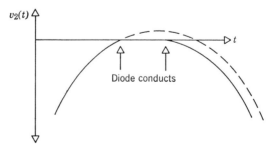

FIGURE 10-14. Detail of the output voltage of a clamping circuit.

256

DIODE LOGIC

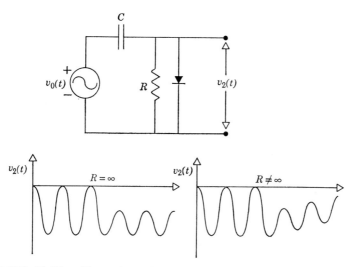

FIGURE 10-15. Clamping circuit with bypass resistor, and output waveforms with and without the resistor.

to the name of *dc restorer* for the circuit, although there is no guarantee that the dc value that the circuit now possesses is anything like that which it had originally, as, for example, before passing through a capacitor.

Another point is that in this circuit, the effect of the reverse leakage current in the diode due to a finite R_r is easily visible. During the full cycle of the sine wave, the capacitor will be discharging through R_r so that the maximum voltage will not quite achieve $-2V_0$, and when the input signal arrives at $+V_0$, the output will try to go slightly positive. At this point the diode conducts again, giving the output wave a flat top (Fig. 10-14).

Note that if the capacitor is small enough so that the time constant R_rC is not much greater than the period of the signal, severe distortion will occur. On the other hand, if R_rC is very large, there will be no easy way for the capacitor to ever discharge, and a reduction of the amplitude of the input wave will result in a situation in which the output is no longer clamped to ground for a long period of time. For this reason, a bypass resistor R is usually placed in parallel with the diode, R being large enough so that distortion is not serious (Fig. 10-15).

DIODE LOGIC

As we stated in the beginning of this chapter, the diode can be most simply thought of as a device with two very different states: on (conducting, $R_f = 0$),

NONLINEAR CIRCUIT ANALYSIS—DIODES

and off (not conducting, $R_r \simeq \infty$). This on-off behavior makes it suited for digital operations, in which the information is not in the form of a linear relationship between two signals, but in a discrete set of states, each different from the other and possessing no overlap in value. In a linear, or analog, system, information could be carried by observing the value of a voltage, say $+5.62 \pm 0.01$ volts. Information could also be carried by the frequency, pulse shape, etc. In a digital system, the information is carried by a sequence of values, which, for a binary system, might be 0 ± 2 volts and 10 ± 2 volts. Any signal within the lower range of values would be called "0" and in the upper range of values "1." To obtain the same information as 5.62 ± 0.01 volts, you would need enough binary pulses to form the binary equivalent of 562 plus a decimal point.

Some of the advantages of digital systems over analog systems are that the quality of the basic devices does not enter as critically into the final result, that only a few simple circuits can perform all possible operations, and that a large number of operations can be performed sequentially without loss of quality or information, since digital signals can be "updated." The last process merely means to set all signals less than the acceptance level (in our case, $+8$ volts) to zero, and all signals between 8 and 12 volts to 10 volts. The pulse shape is also restored (Fig. 10-16).

Finally, the simple circuits required for logical operations are ideally suited to large-scale integrated-circuit techniques (LSI), which means that they may be mass produced cheaply. For these reasons, among others, the digital computer has supplanted the analog computer in almost all applications.

There are two basic types of logical operations: Those that involve the input to a system at a given time, and those that involve both the input at a given time as well as the past history of the device to generate an output. In the first category, we have such circuits as the AND, OR, NOT, and NAND. Each of these titles represents a logical statement. For example, for the AND operation, the statement is: "An output pulse is generated only if an input

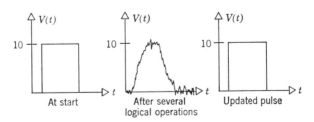

Figure 10-16. Updating a digital pulse.

RECTIFIERS

pulse is present at all inputs simultaneously." An easy way to express this is by way of a *truth table*, which lists all possible combinations of input and output. A 1 indicates that a signal of a given polarity is present, and a 0 indicates that a signal is not present. Beside each basic operation is one possible diode-logic (DL) circuit. For the AND circuit, it can be seen that the diodes are biased *on* if no signal is present, so that only a very small voltage will be present at the output. If one diode is biased *off* due to a positive input pulse, the other will still conduct, so that the output will remain zero. If both are biased off, then the output rises to the value of the smallest input, which gives an output pulse, or a 1 (since we have assumed all input pulses are the same). It should be noted that this common circuit is sometimes called a coincidence circuit.

Operation of the NOT circuit is left as a problem for the student (see problem 4).

The diode-logic circuits shown in Fig. 10-17 are rarely used in this simple form since circuits involving diode-transistor logic (DTL) and transistor-transistor logic (TTL) perform better. Also, note that an AND followed by a NOT with one input at 0 will give a NAND result. Other logical operations are possible with these basic logical units.

The second type of logical operation involves an output that is a function both of the input and the past history of the system. The first of these is the bistable multivibrator (or flip-flop), the heart of scalers, or digital counters. These circuits will be considered in the next chapter.

RECTIFIERS, ZENER DIODES, AND POWER SUPPLIES

Deviating somewhat from our emphasis on fast, low-power electronics, we will discuss a most important application of diodes, namely the establishment of stable, ripple-free voltages for bias purposes. The clipping circuits previously described have the property of taking an alternating voltage and, by removing half of the waveform, generating a periodic voltage that has a nonzero direct-current component under a Fourier decomposition (see Fig. 10-4). By passing this signal through reactive elements that suppress the high-frequency components, we can obtain a direct current $I_{\rm DC} \leq I_{\rm max}/\pi$. However, greater efficiency is obtained when we couple diodes or rectifiers in such a fashion that the whole waveform appears with a single sign.

The diode bridge of Fig. 10-18 results in current flow through a load placed at $v_2(t)$ regardless of the sign of the transformer output, either through

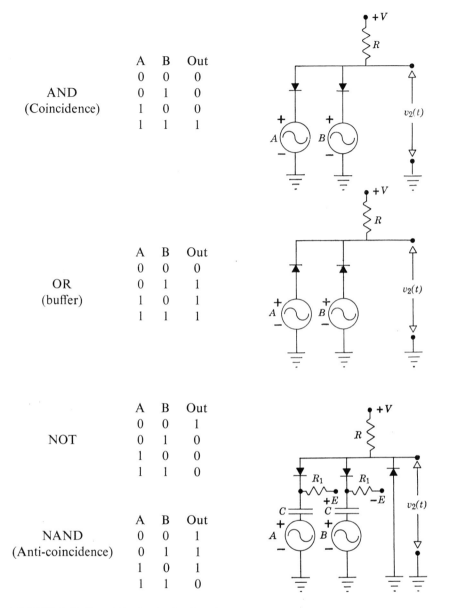

	A	B	Out
AND	0	0	0
(Coincidence)	0	1	0
	1	0	0
	1	1	1

	A	B	Out
OR	0	0	0
(buffer)	0	1	1
	1	0	1
	1	1	1

	A	B	Out
NOT	0	0	1
	0	1	0
	1	0	0
	1	1	0

	A	B	Out
NAND	0	0	1
(Anti-coincidence)	0	1	1
	1	0	1
	1	1	0

FIGURE 10-17. Diode logic for **AND**, **OR**, and **NOT** operations.

RECTIFIERS

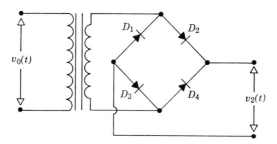

FIGURE 10-18. Full-wave rectifier.

diodes D_1 and D_3 or D_2 and D_4. The output waveform becomes that shown in Fig. 10-19.

The resultant direct current available from such a full-wave rectifier is just twice that from a half-wave rectifier, or

$$I_{DC} \leq \frac{2I_{max}}{\pi}.$$

In practice, the same result is gained through the use of a center-tapped transformer and two diodes (see Problem 10-1).

This output wave form must be carefully filtered and regulated before it is usable as bias on an amplifier, for example. Otherwise, the ac component will also be amplified and added to pre-existing signals. The term *Filtering* is usually applied to the application of passive, reactive elements to selectively suppress certain frequencies in a signal. In the case of power supplies, one wishes to suppress all frequencies >0. Therefore, one places inductors in series and capacitors in parallel to make a section of filter. We will not

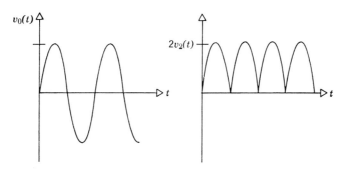

FIGURE 10-19. Input and output waveforms of a full-wave rectifier.

NONLINEAR CIRCUIT ANALYSIS—DIODES

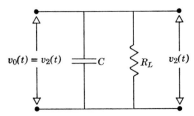

FIGURE 10-20. Capacitive filtering network.

attempt a detailed analysis of passive filtering, since this is available in the texts listed in the Bibliography. However, we shall perform an approximate analysis of capacitive filtering of a full-wave rectified signal. Consider the circuit of Fig. 10-20.

The capacitor charges up to the full voltage $V_{\max}$ through the bridge as the wave goes positive. As soon as the input voltage drops, the diodes in the bridge are back biased and no longer conduct. Therefore, the only current flowing in the load R_L is due to the discharge of the capacitor. We have already solved this series RC circuit, and we have that, during the discharge period,

$$v_2(t) = i_2(t) R_L$$
$$= V_{\max} e^{-t/RC}.$$

The cycle repeats as soon as the input voltage exceeds the capacitor voltage (Fig. 10-21), and the capacitor is raised again to V_0.

The quantity of interest is the amount of ac *ripple* on the dc voltage, which can be defined by

$$\text{ripple} = \frac{\frac{1}{2}(V_0 - V_{\min})}{V_0}.$$

Assume for the moment that the discharge time t_d is approximately equal to the period T of the unrectified wave, so that

$$t_d \simeq T = \frac{1}{f} = \frac{2\pi}{\omega}.$$

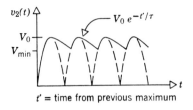

$t' = $ time from previous maximum

FIGURE 10-21. Filtered output of a full-wave rectifier.

RECTIFIERS

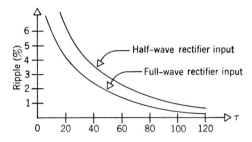

FIGURE 10-22. Ripple as a function of the time constant for half-wave and full-wave rectification.

Then, assuming that the $\tau \gg T$, we can expand the exponent:

$$v_2(t) = V_0 e^{-t/\tau} = V_0\left(1 - \frac{t}{\tau} + \cdots\right)$$

and

$$V_{\min} \simeq V_0\left(1 - \frac{T}{\tau}\right),$$

so that

$$\boxed{\text{ripple} \simeq \frac{T}{2\tau},}$$

where $\tau = R_L C$.

Consider a 120-hertz full-wave rectified output into a 1 K resistive load (Fig. 10-22). Therefore, to obtain $\frac{1}{2}\%$ ripple with a full-wave rectifier input requires an 830 μf capacitor. For a half-wave rectifier, this value must be doubled. These values are quite large.

Filtering can also be accomplished by inductors, and by combinations of resistors, inductors, and capacitors. The table below indicates the amount of dc voltage and ripple for the same time constant.

Filter	Half-wave		Full-wave	
	V_{DC}	Ripple	V_{DC}	Ripple
C, R	$0.32 V_0$	$1/120\tau$	$0.64 V_0$	$1/240\tau$
L, R	$0.32 V_0$	$1/240\tau$	$0.64 V_0$	$1/1130\tau$

263

NONLINEAR CIRCUIT ANALYSIS—DIODES

By adding a capacitor after the inductor in the full-wave case, ripple can be further reduced to about $10^{-6}/LC$.

Assuming that we have obtained a reasonably ripple-free output, the next task will be that of stabilization. While complex systems including feedback and voltage reference sources can be used, we often find that a single *Zener diode* will suffice for many applications. Recall that when a high back bias was applied to a junction diode, a sudden increase of current saw observed at a voltage V_Z, the Zener voltage. What has occurred is that the electric field in the diode is now high enough so that bound electrons can be raised into the conduction state and carry current. These plentiful electrons cause a sharp rise in current; and with a small further increase in voltage, avalanche multiplication occurs, bringing still more bound electrons into the conduction state. The diode is thus manufacturing minority and majority carriers, and if the process were not checked, it would rapidly destroy itself. However, diodes can be specially constructed to take the beating, and these *Zener diodes* can be prepared with V_Z ranging from a few tenths of a volt to around 1000 volts.

Consider the circuit of Fig. 10-23. Suppose that in the Zener region, the diode has an impedance of 30 Ω. Until V_Z is reached, the diode has no regulatory effect, since $R_Z \simeq \infty$. At V_Z, the diode starts to draw current. If the voltage goes from V_Z to $V_Z + 1$ volt, the output voltage v_2 would normally respond linearly. However, the diode starts to conduct like a 30 Ω resistor and, in fact, draws a current of $(V - V_Z)/R_Z = I_Z$. All that is then necessary is to have the additional current flow in the Zener I_Z, through the series resistor R, result in a voltage drop equal to the previous voltage rise. It is left to the student (Problem 10-2) to find the exact relationship, but it should be noted that as $R_L \to \infty$,

$$\Delta V_L \simeq \Delta V_0 \left(\frac{R_Z}{R + R_Z} \right).$$

Note that the Zener will also help reduce ripple. Caution must be exercised not to exceed the power rating of the diode. In addition, the voltage

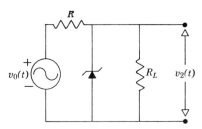

FIGURE 10-23. Zener diode voltage regulator.

TUNNEL DIODES

V_Z is somewhat temperature dependent, and care must be taken to insure that such variations will not adversely affect the system.

TUNNEL DIODES*

The last device that we will discuss is the tunnel diode. This fascinating device was invented in 1958 by Esaki, who was trying to overcome the limitations in switching speed inherent in a *pn* junction. Since there is stored charge at the junction, a reversal of polarity will not immediately turn off a junction which is biased on. Esaki hoped that, by using very high dopings (1 in 10^3 rather than the 1 in 10^8 that is normal), he could thin the diffusion layer and decrease the transition time. The resultant diode characteristics are shown in Fig. 10-24.

This curve was explained as follows: the reverse saturation current I_0 is swamped by the Zener effect even at very low voltages, because the depletion layer has been thinned to approximately 10^{-6} cm from a normal value of 10^{-4} to 10^{-5} cm. This means that the electric field at the junction for the Zener effect is now in tenths of a volt rather than 10 to 100 volts or more. As the junction is forward biased, the depletion layer is further thinned, and the current rises due to transitions between valence and conduction electrons. These transitions are explained by quantum mechanical tunneling through the very thin barrier by electrons at the bottom of the conduction band to the top of the valence band. As the bias is increased, these two bands no longer overlap in energy, and the current drops. Finally, the barrier is decreased to the point that normal diode conduction can take place, and the

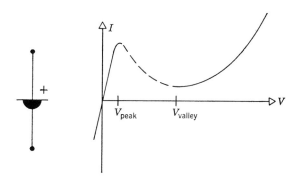

FIGURE 10-24. Tunnel diode schematic and characteristic curve.

---

* Useful references are the *G.E. Tunnel Diode Handbook*, General Electric Company, Schenectady, N.Y., and Millman and Taub, McGraw-Hill (1965), Chapter 12, 452 ff.

NONLINEAR CIRCUIT ANALYSIS—DIODES

current once again rises. The region in which the current drops as the voltage rises corresponds to a region of negative resistance in our piecewise-continuous model.

If we were to try to operate a circuit in this region, we would find that it would be unstable, since a slight *increase* in voltage would give a *decrease* in current. This would result in a *smaller* voltage drop in resistances in the system and a still further increase in voltage. Therefore, the system would switch rapidly into a region in which resistance is positive, along a load line that has been established by the circuit. We will give examples involving tunnel diodes in the next chapter, but in summary we note that

1. Switching times are very fast.
2. Current carrying capabilities are large, since the dopings are high.
3. Voltages are small with $V_p \simeq 0.05$ to 0.15 volt, and $V_v \simeq 0.35$ to 0.50 volt.

One disadvantage of the tunnel diode is that it is a two-terminal device. This can result in design problems, since there is no isolation between the input to the device and the output that results from its operation. A voltage generated in the output circuit is just as capable of switching the device from one state to another as a voltage fed in on the input side. Compare this case to that of a transistor, FET, or tube, in which any voltage present at the output has very small influence on the input in most configurations.

PROBLEMS

1. Design a rectifying bridge for full-wave rectification that uses a center-tapped transformer. Discuss the merits of such an arrangement versus the bridge shown in Fig. 10-18.
2. Solve the circuit shown in Fig. 10-23 and, assuming a linear response in the Zener region (constant R_Z), verify that the approximate formula given in the text is correct.
3. Go back to the ORTEC 220 schematic included as a special project at the end of Chapter 8. Now, analyze the purpose of *all* diodes that occur in that schematic.
4. Explain the operation of the NOT circuit given in the text.
5. Design a coincidence circuit using diodes, capacitors, resistors, and a difference amplifier, and explain its operation. Assume input pulses of equal heights but varying durations and random spacing in time.

EXPERIMENT 12

Experiment 12

Diodes as Clippers and Clampers

I. General

　A. Use dc position of input selector switch on 'scope for all measurements.
　B. The series resistance R in all of the following is not to be set at less than 100 Ω. This resistor limits the current through the diode, and at values less than 100 Ω the diode current can become excessive.
　C. The wiring in this experiment is somewhat more complex than in previous experiments. Use care to ensure that no bare wires or terminals are in accidental contact. Double-check polarities on diodes and power supplies before applying power to any circuit.
　D. When instructed to observe and sketch the waveform, you are to note all pertinent amplitudes and times on the diagram.

Equipment

1. Silicon diodes
2. 2 Decade resistors
3. 20-f capacitor
4. Audio generator
5. 10-volt power supply (dc)
6. Oscilloscope

II. Silicon diode (IN539 or IN540)

　A. Connect circuit as shown.

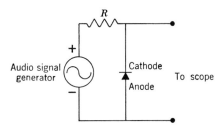

　　1. Use sine function, approximate 500∼/sec on signal generator.
　　2. Set R to 90 KΩ.
　　3. Observe and sketch waveform. Particularly note the amplitude in the negative direction.

NONLINEAR CIRCUIT ANALYSIS—DIODES

B. Reverse the diode and observe and sketch the waveform. Note particularly the amplitude in the positive direction.
  1. Set R to $1000\ \Omega$.
  2. Observe and sketch the waveform.
     Note: There may be an internal capacitor in series with the output of the audio signal generator so that the actual circuit is this:

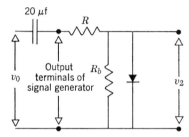

R_b denotes the back resistance of the diode

  3. Switch sig. gen. to square-wave function.
     a. Observe and sketch the waveform.
     b. Switch R to $90\ \mathrm{K}\Omega$ and again observe and sketch.

C. What differences do you find in the waveforms observed in A.3 and B.2? In B.3.a. and B.3.b.? Give an explanation for the observed difference.

D. Connect the circuit as shown.

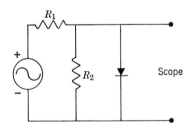

  1. Set $R_1 = 100\ \Omega$;
     $R_2 = 10\ \mathrm{K}\Omega$.
  2. Observe and sketch the waveform with a square-wave input.
  3. Verify the following relation:

$$\frac{A_f}{A_b} = \frac{R'_f}{R_2},$$

EXPERIMENT 12

where $R_f = R_1 + R_f$ and A_f and A_b are the areas under the forward and backward waveforms, respectively. Take $R_f \simeq 15\ \Omega$.

Note that $R_b \gg R_2 \gg R_f$ is still a good approximation.

E. Connect the circuit as shown.

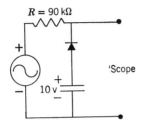

Use a small Transpac 10 v power supply as the "battery," if available. The red terminal on the Transpac is the positive terminal.

1. Set sig. gen. to sine function.
2. Observe and sketch the waveform.
3. Reverse the 10 v power supply, and repeat 2.

TRIGGERS AND MULTIVIBRATORS

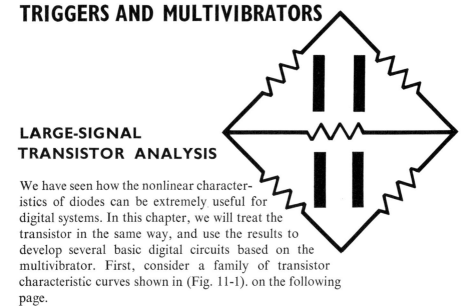

LARGE-SIGNAL TRANSISTOR ANALYSIS

We have seen how the nonlinear characteristics of diodes can be extremely useful for digital systems. In this chapter, we will treat the transistor in the same way, and use the results to develop several basic digital circuits based on the multivibrator. First, consider a family of transistor characteristic curves shown in (Fig. 11-1). on the following page.

We have established a load line in the usual way. In our previous analysis, we were interested only in the linear portion of the transistor surface. At this time, we will consider the extreme portions of the curve. At point S, additional signal in the base of the transistor produces no further increase in collector current. The base has been biased on to the point that both the

TRIGGERS AND MULTIVIBRATORS

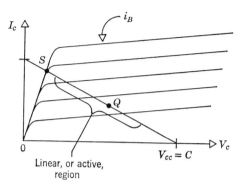

FIGURE 11-1. Transistor characteristic curves and a load line.

BE and *CB* junctions are forward biased, and current is limited only by the total voltage across the transistor and is now independent of base current. The transistor is in *saturation*. At point *C*, on the other hand, the base has been biased so that both junctions are back biased, and only a small reverse saturation current flows. The transistor is in *cut-off*. The active region is the narrow band where the *BE* junction is forward biased and the *CB* junction back biased.

Detailed analysis can do a good job in explaining this large-signal behavior (look up the Ebers-Moll equation in one of the texts in the Bibliography), but in practice, there is a very simple way to describe virtually all transistors. A table can be made of the junction voltages for the various states, depending only on whether the transistor is silicon or germanium.

For an *npn* transistor at room temperature (reverse signs for *pnp*)

	V_{CE} Saturation	V_{BE} Saturation	V_{BE} Active	V_{BE} Cut-off
Silicon	0.3 v	0.7 v	0.7 to 0.5	0.0
Germanium	0.1 v	0.3 v	0.3 to 0.1	−0.1

Note that a slight negative voltage is required to hold a germanium transistor in cut-off. This is due to the relatively large reverse saturation current. Only two other numbers are needed, namely, the dc $\beta (= I_C/I_B)$, and the breakdown voltage across the transistor, called V_{CEO}, collector-to-emitter breakdown voltage. It should be noted here that β_{DC} has a nonzero value at all points on the load line, while β_{AC} is about zero in saturation and cut-off, and assumes a large value only in the active region.

THE BISTABLE MULTIVIBRATOR

MULTIVIBRATORS

We mentioned in the previous chapter that one of the important classes of logical operations involved operations generating an output that was a function both of the input and the past history of the system. A *bistable multivibrator* (*binary*) performs exactly in this manner, since it has two stable and non-overlapping states, and the output depends on which of the two states it is in when a triggering pulse arrives. This is a simple form of memory, since it stores binary information, and if many of these units are arranged so that the output of one provides the input for the next one in the chain, one has a binary scalar. However, the binary is only one member of the important class of multivibrators, the others being the astable multivibrator and the monostable multivibrator, with the Schmitt trigger being a very close relative. Each is basically a two-stage amplifier with strong positive voltage feedback. However, rather than oscillate, the circuit is driven rapidly out of the linear, or small-signal, region into a stable or quasi-stable (long-lived) state in which the transistor, tube, or FET is either in cut-off or saturated. We will choose transistors for all of the analysis, although it should be evident that FETs would work equally well. The members of the multivibrator family are shown in the table.

Device	Description	Uses
(1) Astable Multi	Two unstable states, with the length of time in each one controllable	Oscillators, Timing circuits, Square-wave Generators
(2) Monostable Multi (one-shot, gating circuit)	One stable state, one unstable state of controlled duration	Delay or Gate circuits
(3) Bistable Multi (flip-flop, or binary)	Two stable states	Scalers, Memory, Arithmetic Operations
(4) Schmitt Trigger	One stable state, one state that is maintained only as long as a minimum input is present	Voltage Discriminators, Analog-to-Digital Conversion

THE BISTABLE MULTIVIBRATOR (OR BINARY)

Let us consider the circuit of a transistor binary (Fig. 11-2). From the symmetric nature of the circuit, one might expect that the transistors would

TRIGGERS AND MULTIVIBRATORS

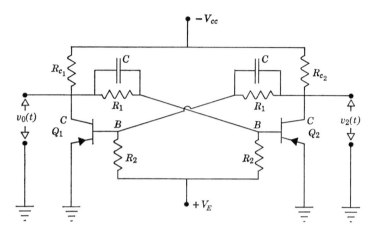

FIGURE 11-2. Bistable multivibrator.

be at identical operating points. Assume that both are operating identically in the active region. Suppose the current I_1 (Fig. 11-3) increases minutely. Then V_{C1} will become somewhat more positive, as will V_{B2}. This is amplified and *inverted* by Q_2; thus, V_{C2} becomes more negative, as does V_{B1}, which in turn is amplified and inverted, so that V_{C1} becomes yet more positive as more current flows. This proceeds until Q_1 is in saturation, and Q_2 in cut-off. At this point, the system is stable. Assume $V_{C1} \simeq V_{B1} \simeq 0$. We can solve for the basic quantities as shown in Fig. 11-3. V_{C1} is about 0 volts. Therefore, we can find the voltage at B:

$$V_{B2} = +1.5\left(\frac{68}{68 + 15}\right) = +1.2 \text{ v.}$$

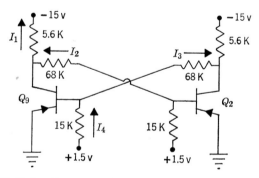

FIGURE 11-3. Dc conditions in a bistable multivibrator.

THE BISTABLE MULTIVIBRATOR

Thus, Q_2 is, indeed in cut-off, since only $+0.1$ volt is required for a Si or Ge. *pnp* transistor. Now

$$I_{C1} = I_1 - I_2;$$

$$I_1 = \frac{15}{5.6 \text{ K}\Omega} = 2.8 \text{ ma};$$

$$I_2 \simeq \frac{1.5}{83 \text{ K}\Omega} \simeq 0.$$

Therefore,
$$I_{C1} = 2.8 \text{ ma}.$$

At this point, we may use the value of β_{DC} to find the current necessary at the base of Q_1 to hold it in saturation. We could also use the characteristic curves generated by a transistor curve tracer if they were available. Assume that $\beta_{DC} = 30$. Then

$$I_{B1} = \frac{2.8}{\beta_{DC}} = 0.09 \text{ ma}.$$

Solving for the collector of Q_2 in the same way as we have done for Q_1,

$$I_3 = \frac{15}{73.6 \text{ K}\Omega} = 0.20 \text{ ma};$$

$$I_4 = \frac{+1.5}{15 \text{ K}\Omega} = +0.10 \text{ ma};$$

$$I_{B1} = I_3 - I_4 = 0.10 \text{ ma, or in saturation}.$$

We have now verified that the circuit is stable with one transistor in cut-off and the other in saturation, and unstable if both are in the active region. To examine details of the transition from one state to another under the impetus of a trigger pulse, we will use an equivalent circuit for the transistor and include a speed-up capacitor in the analysis.

Assume that the transistors are both in the linear region, forced there by a trigger pulse that brings the pair out of cut-off or saturation. The transistor equivalent circuit is shown in Fig. 11-4, where C_i and C_o are the effective input and output capacitance, as before, and we have ignored h_{re}. For the external connections between transistors, we have what looks like a compensated attenuator, going from C_{C1} to T_2 (Fig. 11-5). Compensation occurs for $R_1 C_1 = [r_B R_2/(r_B + R_2)] C_i$. Then the voltage transfer function becomes

$$\bar{T}(s) = \frac{\dfrac{r_B R_2}{r_B + R_2}}{\dfrac{r_B R_2}{r_B + R_2} + R_1},$$

TRIGGERS AND MULTIVIBRATORS

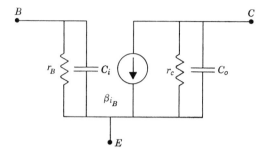

FIGURE 11-4. Transistor equivalent circuit in the active region.

or

$$\bar{T}(s) \simeq \frac{r_B}{R_1}$$

for R_1 and $R_2 \gg r_B$.

Choice of C_1 will then be made so that compensation occurs, resulting in a frequency-independent transfer function. We can now perform a voltage analysis so we have the following single-pole voltage transfer function for a transistor amplifier stage:

$$\bar{T}_1(s) = -g_m r_\| \left(\frac{1}{1 + s\tau_\|}\right);$$

$$\tau_\| = r_\| C_0 \qquad C_0 = C_{CE} + C_{BC};$$

$$\frac{1}{r_\|} = \frac{1}{r_C} + \frac{1}{R_C};$$

$$\bar{T}_1(s) \simeq \frac{-\beta}{r_B C_0}\left(\frac{1}{s + \frac{1}{\tau_\|}}\right).$$

Now proceed around the loop once, but do *not* close the loop yet:

$$\bar{T}(s) = \frac{-\beta_1}{r_{B1} C_{o1}} \left(\frac{1}{s + 1/\tau_\|}\right) \left(\frac{r_{B1}}{R_1}\right) \left(\frac{-\beta_2}{r_{B2} C_{o2}}\right) \left(\frac{1}{s + 1/\tau_\|}\right) \left(\frac{r_{B2}}{R_1}\right).$$

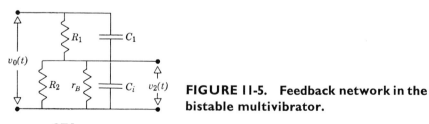

FIGURE 11-5. Feedback network in the bistable multivibrator.

THE BISTABLE MULTIVIBRATOR

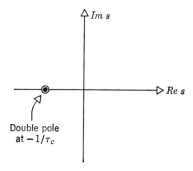

FIGURE 11-6. Pole-zero plot of the transfer function of the bistable multivibrator once the loop has been closed.

For identical transistor impedances $r_{B1} \simeq r_{B2}$, $(\beta_1 \neq \beta_2)$, $C_{o1} = C_{o2}$, and $r_{C1} \simeq r_{C2} = \infty$,

$$\bar{T}(s) = L\left(\frac{1}{1 + s\tau_C}\right)^2,$$

where

$$L = \beta_1\beta_2\left(\frac{R_C}{R_1}\right)^2;$$

$$\tau_C = R_C C_o.$$

This results in a pole-zero plot as shown in Fig. 11-6.

Now close the feedback loop by connecting the output directly to the input, so that

$$\beta_F = 1.$$

Then

$$\bar{T}_f(s) = \frac{\bar{T}(s)}{1 - \beta_F \bar{T}(s)};$$

$$\bar{T}_f(s) = \frac{L}{(1 + s\tau_C)^2 - L}.$$

So the double pole on the negative real axis has been turned into two poles:

$$\alpha_{1,2} = -\frac{1}{\tau_C}(1 \mp \sqrt{L}),$$

or

$$\alpha_{1,2} = -\frac{1}{\tau_C}\left(1 \mp \sqrt{\beta_1\beta_2}\frac{R_C}{R_1}\right).$$

277

TRIGGERS AND MULTIVIBRATORS

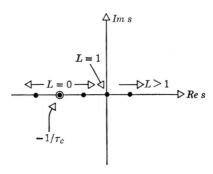

FIGURE 11-7. Pole-zero plot of the transfer function of the bistable multivibrator once the loop has been closed.

As L increases, due to the product $\beta_1\beta_2$ increasing, the double pole splits along the negative real axis (Fig. 11-7). When $L = 1$, one pole reaches the origin. If $L > 1$, one pole is on the *positive* real axis, and corresponds to an exponentially growing signal. The condition for a growing exponential signal is

$$\sqrt{\beta_1\beta_2}\,\frac{R_C}{R_1} \geq 1;$$

$$\beta_1\beta_2 \geq \left(\frac{R_1}{R_C}\right)^2.$$

In the circuit we are using (and will use in the experiment at the end of this chapter),

$$\frac{R_1}{R_C} = 12,$$

so that the threshold for flipping states is

$$\beta_1\beta_2 \geq 144.$$

As we mentioned before, β is almost zero at saturation and cut-off, so that the circuit is stable against minute fluctuations. An input pulse must be large enough to bring the product into this range. Since β is characteristically around 50 to 100, this threshold is reasonably low. Rearrangement of R_1 and R_C could make the circuit either more or less stable.

The output voltage, taken at either collector, is insensitive to many details of the circuit. Almost the full supply voltage appears at the output when the transistor is in cut-off, and only about -0.3 volt appears when the transistor is in saturation. We have already mentioned the applicability of

THE SCHMITT TRIGGER

large-scale integrated-circuit techniques to digital circuits. Binaries and assemblies of binaries are readily available with additional circuitry to aid in presetting, triggering, and resetting (J-K, R-S, etc., binaries). With these improvements, higher transition rates and greater sensitivity can be achieved than with the circuit used in this analysis.

THE SCHMITT TRIGGER

The Schmitt trigger, or emitter-coupled binary, is a very useful modification of the standard binary that is widely used as a voltage discriminator. The basic circuit is sketched in Fig. 11-8.

The circuit acts as a voltage comparator. The input voltage is compared to a voltage V_1 derived from the current through a saturated transistor T_2 and the emitter, or coupling, resistor R_E. The stability of V_1, therefore, depends only on the stability of the bias supply, the temperature dependence of the resistor, and whether T_2 is really saturated. The analysis is very similar to that of the standard binary, the difference being that the base of T_1 is not involved in the switching process; thus, T_2 is normally saturated and T_1 in cut-off. Only when an input voltage exceeds V_1 $(= I_E R_E + V_{BE,\text{active}})$ does T_1 start to amplify. In a similar fashion to the binary, when the loop gain is greater than 1, the circuit regeneratively flips to its second state in which T_2 is in cut-off and T_1 is saturated. At lower loop gains, T_1 merely amplifies the input until finally, as the input voltage rises, T_2 is in cut-off and the output no longer changes. The voltage at the output rises from $(-V_{CC} + I_2 R_{C2})$ to $(-V_{CC})$.

As the input voltage drops, the output voltage retraces the path taken as the voltage rose for loop gains less than or equal to 1. For loop gains greater

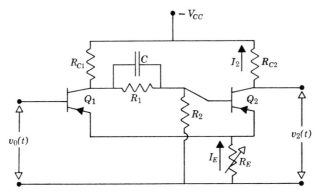

FIGURE 11-8. Schmitt trigger.

TRIGGERS AND MULTIVIBRATORS

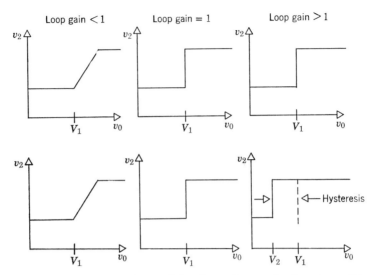

FIGURE 11-9. Output of the Schmitt trigger during transition as a function of loop gain.

than 1, the circuit snaps to its second state at V_1, but as the input drops, the circuit stays in its second state to a voltage V_2 which is less than V_1. This effect is called *hysteresis*. As the loop gain increases, the hysteresis gets worse, while the transition time from one state to the other gets better. The behavior of the circuit for the three choices of loop gain is shown in Fig. 11-9.

For a representative circuit, $V_2 - V_1$ is not small, and may easily be 50% of V_1, or a few volts.

A major use for this circuit is as a voltage discriminator. An example of its operation is shown in Fig. 11-10. Depending on the logic used behind the Schmitt trigger, this circuit can be a *lower-level discriminator* (rejecting all voltages less than V_1) or an *upper-level discriminator* (rejecting all voltages greater than V_1). A lower-level discriminator might perform the updating of digital pulses mentioned in Chapter 10. A pair of these discriminators, one lower level and one upper level, can form a "window" and act as a single-channel analyzer (SCA) with an acceptable voltage range set by the difference between the two V_1's.

In principle, the value of V_1 can be set by adjusting R_E. In practice, small pulses may not be able to drive T_1 out of cut-off; thus, the device will be totally insensitive to a certain range of voltages. For greater sensitivity, the Schmitt trigger is often preceded by a difference amplifier, and the sensitive subtraction mode of this device can supply the needed sensitivity as well as increasing the switching speed for low-voltage pulses.

THE MONOSTABLE MULTIVIBRATOR

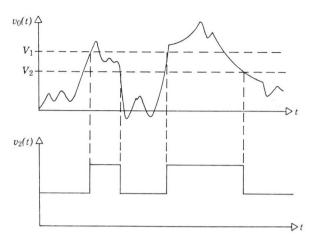

FIGURE 11-10. Schmitt trigger as a lower-level discriminator.

THE MONOSTABLE MULTIVIBRATOR

Consider the circuit of Fig. 11-11. The full name of this circuit is an emitter-coupled monostable multivibrator. Again, we have a coupled pair of transistors, as in the binary or the Schmitt trigger. However, now the feedback loop from the collector of T_1 to the base of T_2 goes via a capacitor, so that no dc path exists to hold T_2 off once it has been triggered. The stable state consists of T_1 in cut-off and T_2 in saturation due to the feedback resistor R_E

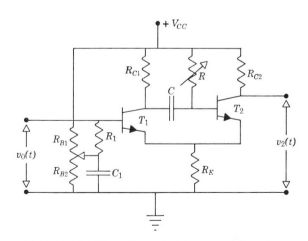

FIGURE 11-11. Monostable multivibrator.

TRIGGERS AND MULTIVIBRATORS

raising the emitter of T_1. Once the circuit has been triggered into its second, or quasi-stable, state, it will remain there only as long as the capacitor can hold the base of T_2 in the cut-off region. This time is set by the charging time of C through R and T_2 (mainly R since T_2 is in cut-off). Although this time can be adjusted by adjusting R, there is another more convenient method that can be used. Analysis shows that the time spent in the quasi-stable state is linearly related to the *supply voltage* V_{CC}. This gives us an electrical method of adjusting a gate width, for example. If the supply voltage is raised beyond a certain value, neither state of the system will be stable, and the circuit will flip back and forth between the two states without any triggering pulse. The multivibrator is then said to be *free running*, and this brings us to the last example of a multivibrator.

THE ASTABLE MULTIVIBRATOR

The circuit shown in Fig. 11-12 represents one form of an astable multivibrator. The analysis of this circuit is very similar to the others, and will be left to the student. Note, however, that no dc path exists for holding either transistor permanently in one state, so that the circuit flips back and forth between the two states with a time set by the RC time constants in the coupling loops. For a symmetrical circuit, the output from either collector is a square wave whose period is

$$T = 1/f = 1.38\tau \qquad \tau = RC.$$

With adjustments of R and C, this period can be varied over a wide range; thus, the circuit provides a high-quality square wave with a frequency set by

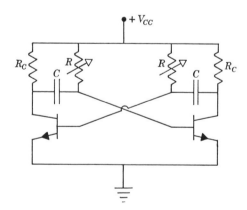

FIGURE 11-12. Astable multivibrator.

TUNNEL DIODE LOGICAL CIRCUITS

passive components. The circuit has many uses in timing and synchronizing systems.

TUNNEL DIODE MULTIVIBRATORS AND DISCRIMINATORS

There are times when an experiment demands the very fastest digital electronics possible, and standard transistor multivibrators may not be adequate. In these cases, the extremely fast switching characteristics of a tunnel diode and the small amount of input power necessary to cause a transition more than overcome the disadvantages of the small operating voltages and the two-terminal nature of the device. Finally, we will discuss tunnel diode-transistor hybrid circuits that possess the advantages of both types of devices. Let us consider the circuit of Fig. 11-13.

We recall from Chapter 10 that the current-voltage characteristics of a tunnel diode possess a region of negative resistance, as indicated in Fig. 11-14. By adjusting the bias supply E_b, we can move the load line around until it intersects the tunnel diode characteristic curve at the point indicated on the diagram as Q. Now assume that a positive pulse arrives in the circuit. It is equivalent to suddenly raising E_b by the voltage of the incoming pulse. If this new "load line" moves point Q over the peak of the curve, the voltage will jump discontinuously to point A, and then follow the tunnel diode curve to the new stable point Q'. The time to reach this point would be roughly proportional to the inductance divided by the effective resistance in the circuit, R and R_{diode}. However, we will assume that, before the process has gone that far, the initial pulse has gone away, so that no stable point Q' exists, and the current drops until the point B is reached. Here the circuit again jumps discontinuously to point C, and then decays back to Q. A piecewise-continuous

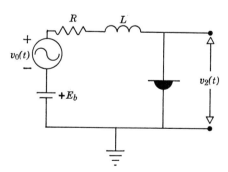

FIGURE 11-13. Tunnel diode monostable multivibrator.

TRIGGERS AND MULTIVIBRATORS

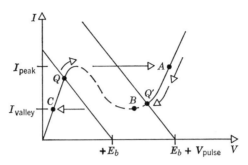

FIGURE 11-14. Tunnel diode characteristic curve and two possible load lines.

model can be used to analyze this process,* and the results are that, for reasonable values for the tunnel diode and using perhaps 50 ohms for R,

$$T \approx 0.01L.$$

Thus, 100 microhenries would give about 1 microsecond for the gate duration. The inherent switching time is in nanoseconds, so that the observed switching time will be strongly dependent on any stray capacitance present.

We have seen how we can make a monostable multivibrator, using a single tunnel diode, and take advantage of the inherent switching speed of the device. Examination of the tunnel diode characteristic curve shows that we just as easily have picked a load line that intersected the curve at two stable points or no stable points (in the dashed region of the curve). In the first case, we would have a bistable multivibrator, and in the second, an astable multivibrator. The bistable circuit is especially useful, since the fast transition time allows very fast count rates, such as 100 megahertz. Thus, tunnel diode scalers are very fast and can provide an input for slower transistor units operating in the 1- to 10-megahertz region.

Finally, we should mention tunnel diode-transistor hybrid circuits. The problem with switching transistors is that a large amount of charge has to be moved around, and base currents in the microampere region are required. If one wishes fast switching, the charge has to be supplied quickly, and the current requirements go up. The tunnel diode has the sensitivity, but it lacks much of an output voltage, characteristically in the tenths-of-a-volt range. If one uses the tunnel diode to supply the base with a fast signal, the result is called a tunnel diode-transistor hybrid, and one can draw characteristic

---

* Millman and Taub, p. 489.

TUNNEL DIODE LOGICAL CIRCUITS

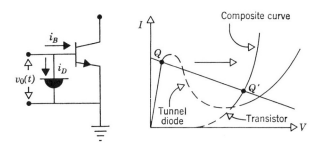

FIGURE 11-15. Tunnel diode-transistor hybrid, characteristic curves, and a load line.

current-voltage curves for the hybrid that still retain the tunnel diode's negative resistance behavior.

If we combine the characteristics of each device on a single graph, the result is shown in Fig. 11-15. We have drawn a load line such that the stable point Q lies just below the low-voltage current peak. A small input voltage will trigger the tunnel diode to switch to point Q', at which point the transistor will be in a region in which it is biased on strongly. No current will be flowing yet in the transistor, but that does not matter, since the response has taken place and the transistor will "catch up" with the tunnel diode in due time. The circuit shown in Fig. 11-16 indicates a practical way of combining the devices to produce a circuit that can act as a discriminator, astable, monostable, or bistable multivibrator.

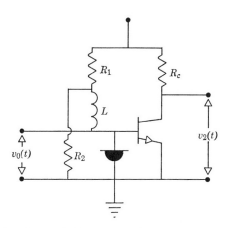

FIGURE 11-16. Tunnel diode-transistor hybrid multivibrator.

TRIGGERS AND MULTIVIBRATORS

Experiment 13

The Transistor Binary

I. Circuit

    A. Wire the circuit as shown.
    B. For the positive pulse source, a pulse generator is to be used.

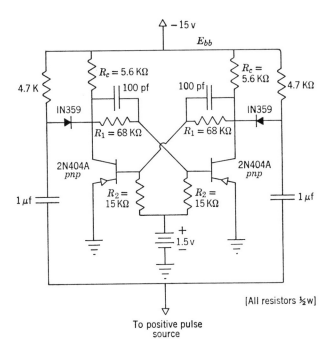

II. Dc measurements

    A. Using the dc VTVM, measure the voltages at E_{bb}, both collectors, and both bases. Verify that one transistor is in saturation, and the other is in cut-off.
    B. The IN539's are known as *steering diodes*. What is the "bias" across each? How does their use permit switching of the binary from either stable state to the other from the same signal source and with the same signal polarity?

EXPERIMENT 13

III. Ac measurements

A. Connect the HP pulse generator into the circuit.
B. Trigger the 'scope from the pulse generator, and set the amplitude of the input pulse to the binary to approximately 5–10 v. Set the pulse width to ~0.2 µsec.
C. Observe the waveform at either collector. Sketch the waveform, showing the position of the input pulses in relation to the output. Note the frequency division; i.e., two input pulses for one output pulse.

   Note: If the 'scope appears to have two horizontal traces, the binary is not triggering reliably. Adjust the Pulse Rate control for a stable pattern. Do not set the Delay to more than 10 µsec.

D. Vary the height and width of the input pulse systematically, noting any changes in the output of the binary. Especially note the minimum pulse width and pulse height at which the binary will trigger stably.

BIBLIOGRAPHY

As we stated in the preface, we have deliberately limited this text to a content that, we feel, can be handled in a semester or, with some deletions, one quarter. Thus, we have concentrated on concepts in circuit analysis and have left out almost all physical description of devices such as tubes, transistors, FETs and integrated circuits, as well as many fascinating and useful circuits. We therefore note here some texts that we have found useful for supplementary study. We especially recommend some of the easier texts if the student has not had introductory electronics as part of a freshman or sophomore course in electricity and magnetism. The texts are listed roughly in order of increasing difficulty.

Halliday, D. and R. Resnick, *Physics, Part II*. John Wiley & Sons (1962).
  Good introduction to the physics involved and to simple circuits.

Phillips, L. F. *Electronics for Experimenters in Chemistry, Physics, and Biology*. John Wiley & Sons (1966). An inexpensive text that describes both devices and circuits for the student with limited background.

Malmstadt, H. V. Enke G. G., and Toren, E. C. *Electronics for Scientists*. (Benjamin1963).
  A large and discursive text that provides abundant material on a moderate level of difficulty. There are many circuits of interest to the scientist in this text and in its matching volume, *Digital Electronics for Scientists*., W. A. Benjamin (1969)

Gray, P. E. and C. L. Searle, *Electronic Principles*. John Wiley & Sons (1967).

Millman, J. and Taub, H. *Pulse, Digital, and Switching Waveforms*. McGraw-Hill (1965).
  Two extensive texts designed for electrical engineers that will prove very useful, especially in the laboratory. Many excellent circuits and relatively sophisticated analysis.

Owen and Keaton, *Fundamentals of Electronics*. Harper & Row (3 volumes).
  Clear, detailed, and rigorous treatment using Laplace transforms for all analysis. Probably considerably more detailed than required for the average scientist.

For the mathematics, many texts are suitable, but we like

Sokolnikoff, I. S. and R. M. Redheffer, *Mathematics of Physics and Modern Engineering*. McGraw-Hill (1966).

INDEX

Active delay line differentiator, 217
Admittance, 33
    admittance matrix, 123
Alpha, 174
Amplification, 107
    amplification factor, 119
Amplifiers, cascaded, 204
    current, 171
    difference, 196
    Gaussian, 205
    logarithmic, 220
    operational, 209
    rush current, 199
    two stage feedback, 156, 165
    voltage, 107
Astable multivibrator, 282
Attenuators, compensated, 22, 52
    overcompensation, 25
    uncompensated, 17
    undercompensation, 25

Bandpass, 132
Bandwidth and noise, 239
Base, transistor, 173
Beta, transistor, 175
Biasing, self, 120
Bias conditions, 176
Bistable multivibrator, 273
Blocking circuit, 9
Buffer, 152

Capacitance, interelectrode capacitance, 3, 121
Capacitive reactance, 32
Cascaded amplifiers, 204
CAT, 244
Cathode, 110
Cathode follower, 152
Characteristic impedances, 79

Charge sensitive preamplifier, 215
Charging currents, 114
Circuits, AND, 260
    blocking, 9
    cathode follower, 152
    clamping, 255
    clipping, 86, 251
    Darlington, 192
    differentiating, 5, 47, 51, 213
    DC restorer, 256
    high pass, 5
    integrating, 9
    limiting, 251
    low pass, 16
    NAND, NOT, OR, 260
    RC, 5, 40, 51
    RCL, 3, 26, 30, 54
    RL, 40, 51
    tank, 28
    voltage doubler, 256
Collector, 173
Common base mode, 184
Common emitter mode, 180
Common mode rejection ratio, 198
Compensation condition, 24
Complex admittance, 33
Complex frequency, 46
Complex impedance, 32
Compensation condition, 24, 54
    overcompensation, 25
    undercompensation, 25
Conductance, 33
Conservation laws, energy, 3
    charge, 3
Conventional current, 110
Conversion formulas, common emitter
    to common base, 188
    common emitter to common
        collector, 188

289

INDEX

Critically damped RCL circuit, 30, 56
Current feedback, 148
Current transfer function, 68, 182, 186
Currents, charging, 114
  loop, 18
  real, 18
  transfer, 114
Cut-off frequency, transistor, common emitter, 183
  common base, 186

Damping, 55
Darlington circuit, 192
DC circuit scheme, 50
Decibel, 16
Delay lines, 75
Delay line clipping, single and double, 86
Delay time, 11, 205
Delayed differentiator, 87
Depletion region, 172
Difference amplifier, 196
Differentiating circuits, 5, 9, 47, 51, 213, 217
Diffusion line, 92
Diodes, junction, 248
  thermionic, 248
  tunnel, 265
Diode logic, 257
Discriminator, 279
Drain, 111
Droop, 131

Emitter, 173
Emitter-collector current gain, 186
Energy, 3
Equivalent circuits, 20, 37, 65, 133, 153

Feedback, 147
  feedback factor, 150
  negative voltage feedback, 149
  positive voltage feedback, 159
FET, 111
Filtering, 261
Fourier analysis, 26
Frequency, 6

Gain, 16, 51, 119, 163, 186, 194, 205
Gain-bandwidth product, rush current amplifier, 202
  transistor amplifier, 183
  triode or FET amplifier, 133

Gate, 111
Gaussian amplifier, 205
Generalized impedances, 50
Grid, 110
  bias, 117
  leak resistance, 128
  resistance, 117

Half power points, lower, 15
  of an amplifier, 132
  upper, 17
Holes, 111
Hysteresis, 280

Impedance, 37, 65
  characteristic impedance, 76
  transfer impedance, 68
Impedance matcher, 152
Inductance, 3
  mutual inductance, 58
Inductive reactance, 32
Input impedance, 37, 65
  amplifier, 108
  cathode follower, 155
  Darlington, 195
  transformer, 61
  triode, 126
Interelectrode capacitance, 121
Integrating circuits, 9
  op-amp or Miller integrator, 213

Johnson noise, 236
Junction diode, 248
Junction, PN, 173

Kirchhoff's laws, 3

Laplace transforms, 45
  table, 66
Large signal transistor analysis, 271
Linear device, 110
Linear integrated circuits, 219
Lissajou figures, 72
Load line, 118
Lock-in amplifier, 244
Logarithmic amplifier, 220
Logarithmic decrement, 28
Low pass circuit, 16

Majority carriers, 111
Method of undetermined coefficients, 6

INDEX

Mid frequency gain, 108
Miller effect, 127
Miller integrator, 213
Minority carriers, 111
Models, large signal transistor, 271
  lumped parameter, 2
  piecewise continuous, 250
  small signal transistor, 110
Multistage amplifiers, 127
Multivibrators, 273
  astable, 282
  bistable, 273
  monostable, 281
  Schmitt trigger, 279
  tunnel diode, 283
Mutual inductance, 58

Negative resistance, 135
NMR probe, 97
Nodes, 18
Noise, 235
Non-inductive transmission line, 91
Norton's theorem, 37

Ohm's law, 4, 32
One-over-f noise, 238
Operating point, 115
Operational amplifiers, 209
  adder, 214
  differentiator, 213
  integrator, 213
  inverter, 212
  transfer function, 211
Oscillator, 28, 160
Oscilloscope probe, 18
Output impedance, 37, 65
  amplifier stage, 134
  cathode follower, 154
  Darlington, 194
  feedback amplifier, 151
  transformer, 61
Overcompensation, 25
Overdamped RCL circuit, 28

Partial fractions, 53
Pentode, 135, 165
Phase, 6
Phase shift, 15, 17, 52
Phasors, 31
Pinch-off, 112
Plate, 110

Plate (*Continued*)
  characteristics, 114
  resistance, 116
PN junction, 173
Pole, 46
Pole trajectories, 57, 157, 161
Pole-zero cancellation, 54
Pole-zero plot, 46
Power dissipation hyperbola, 118
Preamplifier, 215
Probe, 18
Pulse width, 7

Q of a circuit, 28, 33, 35

Reactance, 32
Rectangular pulse, 7
Rectifier, full-wave, 261
  half-wave, 259
Reflection, 80
Resistance, 3
Resonant condition, 26
RF circuitry, 63
RF transmission lines, 94
Ringing, 55
Ripple, 262
Rise time, 11
  of cascaded amplifiers, 205
RLC circuit, 26
Rush transistor current amplifier, 199

Saturation, 272
Screen grid, 135
Selectivity, 34
Selfbiasing, 120
Shifting theorem, 49
Short circuit theorem, 37
Shot noise, 237
Signal to noise ratio, 238
Sinusoidal input, DC circuit scheme, 51
  RC and RL circuits, 13
  RCL circuits, 30
Small signal model, 110
Source, 111
Special project in op amp analysis, 223
Standing wave ratio, 100
Steady state solution, 14
Step input, 6
Superposition theorem, 37
Susceptance, 33

291

INDEX

Tank circuit, 28
Tetrode, 135
Three dB points, 16
Thevenin's theorem, 37
Time constant, RC, 7
  RL, 51
Transconductance, 116
Transfer characteristics, 115
Transfer currents, 114
Transfer function, 48, 51, 64
Transformers, 57
  ideal, 59
  impedances, 61
Transient solution, 14
Transistors, 172
  bias conditions, 176
  capacitance, 174
  H-equations, 175
  stabilization, 176
  surface, 174
Transmission lines, 75
  attenuation, 90
  characteristic impedance, 79
  diffusion line, 92
  half-wavelength line, 96
  loss-less, 75
  non-inductive, 91
  multiple reflections, 88
  quarter-wavelength line, 96
  reflection coefficients, 80

Transistors (*Continued*)
  termination, 79
  transit time, 78
  voltage attenuation ratio, 80
Triode, 110
Triode equations, 115, 123
Tube surface, 114
Tunnel diodes, 265
  discriminators, 283
  hybrid, 284
  multivibrators, 283

Uncompensated attenuator, 17
Undercompensation, 25
Underdamped RCL circuit, 29
Unit step function, 6

Vacuum tube, 110
Virtual ground, 211
Voltage amplifier, 108
Voltage attenuation ratio, 80
Voltage divider, 50
Voltage drops, 4
Voltage feedback, 148

Wave equation, 77
Wien bridge oscillator, 160
Wiring capacitance, 131

Zener diode, 264